Geological History of Earth

Compiled by
Janelle Pounds

Scribbles

Year of Publication 2018

ISBN : 9789352979257

Book Published by

Scribbles

(An Imprint of Alpha Editions)

email - alphaedis@gmail.com

Produced by: PediaPress GmbH
Limburg an der Lahn
Germany
http://pediapress.com/

Contents

Appendix   335

Article Licenses   361

Index   363

Precambrian

Geological history of Earth

The **geological history of Earth** follows the major events in Earth's past based on the geologic time scale, a system of chronological measurement based on the study of the planet's rock layers (stratigraphy). Earth formed about 4.54 billion years ago by accretion from the solar nebula, a disk-shaped mass of dust and gas left over from the formation of the Sun, which also created the rest of the Solar System.

Earth was initially molten due to extreme volcanism and frequent collisions with other bodies. Eventually, the outer layer of the planet cooled to form a solid crust when water began accumulating in the atmosphere. The Moon formed soon afterwards, possibly as a result of the impact of a planetoid with the Earth. Outgassing and volcanic activity produced the primordial atmosphere. Condensing water vapor, augmented by ice delivered from comets, produced the oceans.

As the surface continually reshaped itself over hundreds of millions of years, continents formed and broke apart. They migrated across the surface, occasionally combining to form a supercontinent. Roughly 750[1] million years ago, the earliest-known supercontinent Rodinia, began to break apart. The continents later recombined to form Pannotia, 600 to 540[2] million years ago, then finally Pangaea, which broke apart 200[3] million years ago.

The present pattern of ice ages began about 40[4] million years ago, then intensified at the end of the Pliocene. The polar regions have since undergone repeated cycles of glaciation and thaw, repeating every 40,000–100,000 years. The last glacial period of the current ice age ended about 10,000 years ago.

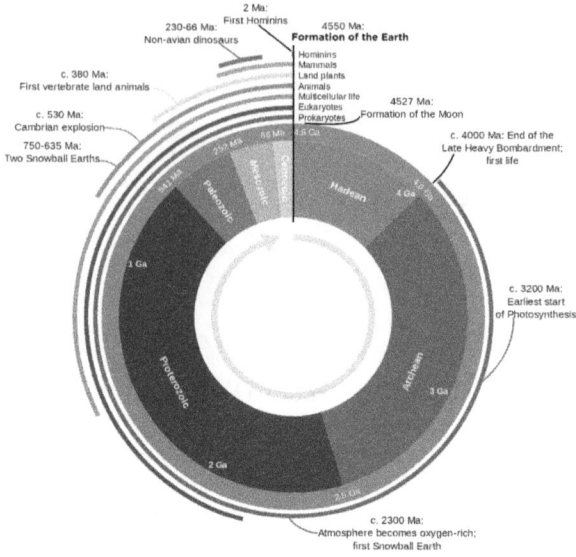

Figure 1: *Geologic time represented in a diagram called a geological clock, showing the relative lengths of the eons of Earth's history and noting major events*

Precambrian

Life timeline

Figure 2: *Artist's conception of a protoplanetary disc*

Axis scale: million years

Also see: *Human timeline* and *Nature timeline*

The Precambrian includes approximately 90% of geologic time. It extends from 4.6 billion years ago to the beginning of the Cambrian Period (about 541 Ma). It includes three eons, the Hadean, Archean, and Proterozoic.

Major volcanic events altering the Earth's environment and causing extinctions may have occurred 10 times in the past 3 billion years.

Hadean Eon

During Hadean time (4.6–4 Ga), the Solar System was forming, probably within a large cloud of gas and dust around the sun, called an accretion disc from which Earth formed 4,500[5] million years ago. The Hadean Eon is not formally recognized, but it essentially marks the era before we have adequate record of significant solid rocks. The oldest dated zircons date from about 4,400[6] million years ago.

Earth was initially molten due to extreme volcanism and frequent collisions with other bodies. Eventually, the outer layer of the planet cooled to form a solid crust when water began accumulating in the atmosphere. The Moon formed soon afterwards, possibly as a result of the impact of a large planetoid with the Earth. Some of this object's mass merged with the Earth, significantly altering its internal composition, and a portion was ejected into space. Some of the material survived to form an orbiting moon. More recent potassium isotopic studies suggest that the Moon was formed by a smaller, high-energy, high-angular-momentum giant impact cleaving off a significant portion of the Earth. Outgassing and volcanic activity produced the primordial atmosphere. Condensing water vapor, augmented by ice delivered from comets, produced the oceans.

During the Hadean the Late Heavy Bombardment occurred (approximately 4,100 to 3,800[7] million years ago) during which a large number of impact craters are believed to have formed on the Moon, and by inference on Earth, Mercury, Venus and Mars as well.

Archean Eon

The Earth of the early Archean (4,000 to 2,500[8] million years ago) may have had a different tectonic style. During this time, the Earth's crust cooled enough that rocks and continental plates began to form. Some scientists think because the Earth was hotter, that plate tectonic activity was more vigorous than it is today, resulting in a much greater rate of recycling of crustal material. This may have prevented cratonisation and continent formation until the mantle cooled and convection slowed down. Others argue that the subcontinental lithospheric mantle is too buoyant to subduct and that the lack of Archean rocks is a function of erosion and subsequent tectonic events.

In contrast to the Proterozoic, Archean rocks are often heavily metamorphized deep-water sediments, such as graywackes, mudstones, volcanic sediments and banded iron formations. Greenstone belts are typical Archean forma-tions, consisting of alternating high- and low-grade metamorphic rocks. The high-grade rocks were derived from volcanic island arcs, while the low-grade metamorphic rocks represent deep-sea sediments eroded from the neighbor-ing island rocks and deposited in a forearc basin. In short, greenstone belts represent sutured protocontinents.

The Earth's magnetic field was established 3.5 billion years ago. The solar wind flux was about 100 times the value of the modern Sun, so the presence of the magnetic field helped prevent the planet's atmosphere from being stripped away, which is what probably happened to the atmosphere of Mars. However, the field strength was lower than at present and the magnetosphere was about half the modern radius.

Proterozoic Eon

The geologic record of the Proterozoic (2,500 to 541[9] million years ago) is more complete than that for the preceding Archean. In contrast to the deep-water deposits of the Archean, the Proterozoic features many strata that were laid down in extensive shallow epicontinental seas; furthermore, many of these rocks are less metamorphosed than Archean-age ones, and plenty are unaltered. Study of these rocks show that the eon featured massive, rapid continental accretion (unique to the Proterozoic), supercontinent cycles, and wholly modern orogenic activity. Roughly 750^1 million years ago,[10] the earliest-known supercontinent Rodinia, began to break apart. The continents later recombined to form Pannotia, 600–540 Ma.

The first-known glaciations occurred during the Proterozoic, one began shortly after the beginning of the eon, while there were at least four during the Neoproterozoic, climaxing with the Snowball Earth of the Varangian glaciation.

Phanerozoic Eon

The **Phanerozoic** Eon is the current eon in the geologic timescale. It covers roughly 541 million years. During this period continents drifted about, eventually collected into a single landmass known as Pangea and then split up into the current continental landmasses.

The Phanerozoic is divided into three eras – the Paleozoic, the Mesozoic and the Cenozoic.

Most of biological evolution occurred during this time period.

Paleozoic Era

The **Paleozoic** spanned from roughly 541 to 252^{11} million years ago (Ma) and is subdivided into six geologic periods; from oldest to youngest they are the Cambrian, Ordovician, Silurian, Devonian, Carboniferous and Permian. Geologically, the Paleozoic starts shortly after the breakup of a supercontinent called Pannotia and at the end of a global ice age. Throughout the early Paleozoic, the Earth's landmass was broken up into a substantial number of relatively small continents. Toward the end of the era the continents gathered together into a supercontinent called Pangaea, which included most of the Earth's land area.

Cambrian Period

The **Cambrian** is a major division of the geologic timescale that begins about 541.0 ± 1.0 Ma. Cambrian continents are thought to have resulted from the breakup of a Neoproterozoic supercontinent called Pannotia. The waters of the Cambrian period appear to have been widespread and shallow. Continental drift rates may have been anomalously high. Laurentia, Baltica and Siberia remained independent continents following the break-up of the supercontinent of Pannotia. Gondwana started to drift toward the South Pole. Panthalassa covered most of the southern hemisphere, and minor oceans included the Proto-Tethys Ocean, Iapetus Ocean and Khanty Ocean.

Ordovician period

The **Ordovician** period started at a major extinction event called the Cambrian–Ordovician extinction event some time about 485.4 ± 1.9 Ma. During the Ordovician the southern continents were collected into a single continent called Gondwana. Gondwana started the period in the equatorial latitudes and, as the period progressed, drifted toward the South Pole. Early in the Ordovician the continents Laurentia, Siberia and Baltica were still independent continents (since the break-up of the supercontinent Pannotia earlier), but Baltica began to move toward Laurentia later in the period, causing the Iapetus Ocean to shrink between them. Also, Avalonia broke free from Gondwana and began to head north toward Laurentia. The Rheic Ocean was formed as a result of this. By the end of the period, Gondwana had neared or approached the pole and was largely glaciated.

The Ordovician came to a close in a series of extinction events that, taken together, comprise the second-largest of the five major extinction events in Earth's history in terms of percentage of genera that became extinct. The only larger one was the Permian-Triassic extinction event. The extinctions occurred approximately 447 to 444[12] million years ago and mark the boundary between the Ordovician and the following Silurian Period.

The most-commonly accepted theory is that these events were triggered by the onset of an ice age, in the Hirnantian faunal stage that ended the long, stable greenhouse conditions typical of the Ordovician. The ice age was probably not as long-lasting as once thought; study of oxygen isotopes in fossil brachiopods shows that it was probably no longer than 0.5 to 1.5 million years. The event was preceded by a fall in atmospheric carbon dioxide (from 7000ppm to 4400ppm) which selectively affected the shallow seas where most organisms lived. As the southern supercontinent Gondwana drifted over the South Pole, ice caps formed on it. Evidence of these ice caps have been detected in Upper Ordovician rock strata of North Africa and then-adjacent northeastern South America, which were south-polar locations at the time.

Silurian Period

The **Silurian** is a major division of the geologic timescale that started about 443.8 ± 1.5 Ma. During the Silurian, Gondwana continued a slow southward drift to high southern latitudes, but there is evidence that the Silurian ice caps were less extensive than those of the late Ordovician glaciation. The melting of ice caps and glaciers contributed to a rise in sea levels, recognizable from the fact that Silurian sediments overlie eroded Ordovician sediments, forming an unconformity. Other cratons and continent fragments drifted together near the equator, starting the formation of a second supercontinent known as Euramerica. The vast ocean of Panthalassa covered most of the northern hemisphere.

Other minor oceans include Proto-Tethys, Paleo-Tethys, Rheic Ocean, a sea-way of Iapetus Ocean (now in between Avalonia and Laurentia), and newly formed Ural Ocean.

Devonian Period

The **Devonian** spanned roughly from 419 to 359 Ma. The period was a time of great tectonic activity, as Laurasia and Gondwana drew closer together. The continent Euramerica (or Laurussia) was created in the early Devonian by the collision of Laurentia and Baltica, which rotated into the natural dry zone along the Tropic of Capricorn. In these near-deserts, the Old Red Sandstone sedimentary beds formed, made red by the oxidized iron (hematite) characteristic of drought conditions. Near the equator Pangaea began to consolidate from the plates containing North America and Europe, further raising the northern Appalachian Mountains and forming the Caledonian Mountains in Great Britain and Scandinavia. The southern continents remained tied together in the supercontinent of Gondwana. The remainder of modern Eurasia lay in the Northern Hemisphere. Sea levels were high worldwide, and much of the land lay submerged under shallow seas. The deep, enormous Panthalassa (the "universal ocean") covered the rest of the planet. Other minor oceans were Paleo-Tethys, Proto-Tethys, Rheic Ocean and Ural Ocean (which was closed during the collision with Siberia and Baltica).

Carboniferous Period

The **Carboniferous** extends from about 358.9 ± 0.4 to about 298.9 ± 0.15 Ma.

A global drop in sea level at the end of the Devonian reversed early in the Carboniferous; this created the widespread epicontinental seas and carbonate deposition of the Mississippian. There was also a drop in south polar temperatures; southern Gondwana was glaciated throughout the period, though it is uncertain if the ice sheets were a holdover from the Devonian or not. These conditions apparently had little effect in the deep tropics, where lush coal swamps flourished within 30 degrees of the northernmost glaciers. A mid-Carboniferous drop in sea-level precipitated a major marine extinction, one that hit crinoids and ammonites especially hard. This sea-level drop and the associated unconformity in North America separate the Mississippian Period from the Pennsylvanian period.

The Carboniferous was a time of active mountain building, as the supercontinent Pangea came together. The southern continents remained tied together in the supercontinent Gondwana, which collided with North America-Europe (Laurussia) along the present line of eastern North America. This continental collision resulted in the Hercynian orogeny in Europe, and the Alleghenian orogeny in North America; it also extended the newly uplifted Appalachians

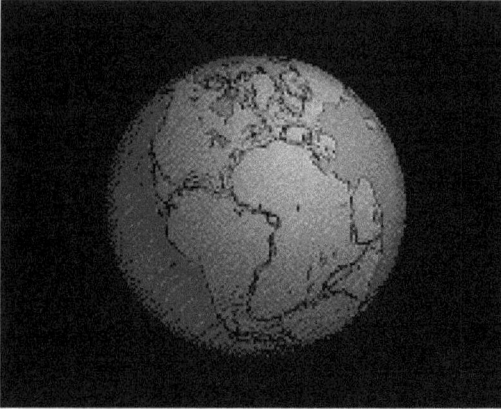

Figure 3: *Pangaea separation animation*

southwestward as the Ouachita Mountains. In the same time frame, much of present eastern Eurasian plate welded itself to Europe along the line of the Ural mountains. There were two major oceans in the Carboniferous the Panthalassa and Paleo-Tethys. Other minor oceans were shrinking and eventually closed the Rheic Ocean (closed by the assembly of South and North America), the small, shallow Ural Ocean (which was closed by the collision of Baltica, and Siberia continents, creating the Ural Mountains) and Proto-Tethys Ocean.

Permian Period

The **Permian** extends from about 298.9 ± 0.15 to 252.17 ± 0.06 Ma.

During the Permian all the Earth's major land masses, except portions of East Asia, were collected into a single supercontinent known as Pangaea. Pangaea straddled the equator and extended toward the poles, with a corresponding effect on ocean currents in the single great ocean (*Panthalassa*, the *universal sea*), and the Paleo-Tethys Ocean, a large ocean that was between Asia and Gondwana. The Cimmeria continent rifted away from Gondwana and drifted north to Laurasia, causing the Paleo-Tethys to shrink. A new ocean was growing on its southern end, the Tethys Ocean, an ocean that would dominate much of the Mesozoic Era. Large continental landmasses create climates with extreme variations of heat and cold ("continental climate") and monsoon conditions with highly seasonal rainfall patterns. Deserts seem to have been widespread on Pangaea.

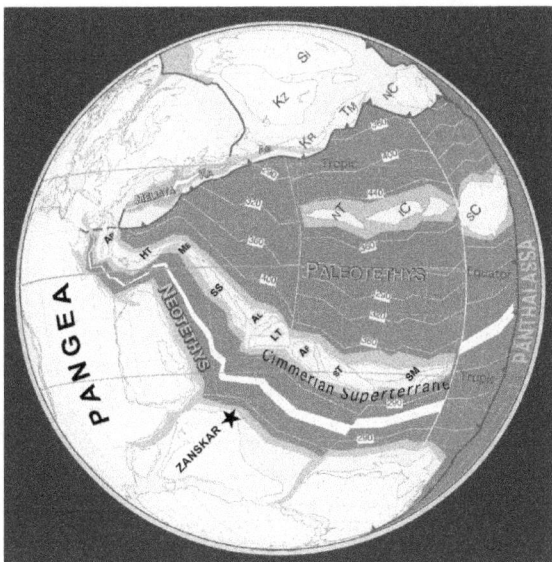

Figure 4: *Plate tectonics- 249[13] million years ago*

Mesozoic Era

The **Mesozoic** extended roughly from 252 to 66[15] million years ago.

After the vigorous convergent plate mountain-building of the late Paleozoic, Mesozoic tectonic deformation was comparatively mild. Nevertheless, the era featured the dramatic rifting of the supercontinent Pangaea. Pangaea gradually split into a northern continent, Laurasia, and a southern continent, Gondwana. This created the passive continental margin that characterizes most of the Atlantic coastline (such as along the U.S. East Coast) today.

Triassic Period

The **Triassic** Period extends from about 252.17 ± 0.06 to 201.3 ± 0.2 Ma. During the Triassic, almost all the Earth's land mass was concentrated into a single supercontinent centered more or less on the equator, called Pangaea ("all the land"). This took the form of a giant "Pac-Man" with an east-facing "mouth" constituting the Tethys sea, a vast gulf that opened farther westward in the mid-Triassic, at the expense of the shrinking Paleo-Tethys Ocean, an ocean that existed during the Paleozoic.

The remainder was the world-ocean known as Panthalassa ("all the sea"). All the deep-ocean sediments laid down during the Triassic have disappeared

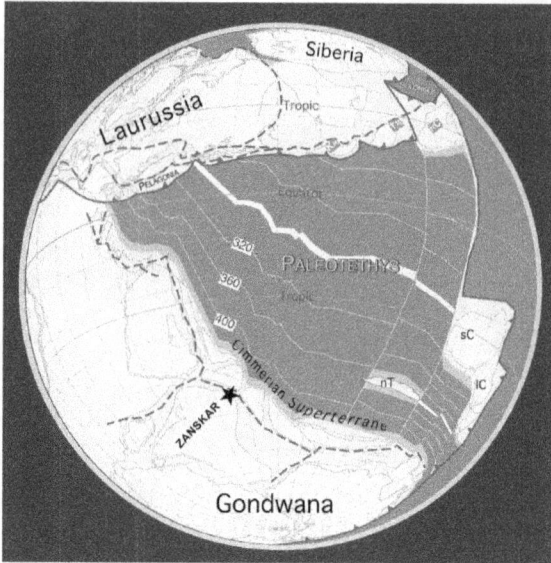

Figure 5: *Plate tectonics- 290[14] million years ago*

through subduction of oceanic plates; thus, very little is known of the Triassic open ocean. The supercontinent Pangaea was rifting during the Triassic—especially late in the period—but had not yet separated. The first nonmarine sediments in the rift that marks the initial break-up of Pangea—which separated New Jersey from Morocco—are of Late Triassic age; in the U.S., these
thick sediments comprise the Newark Supergroup. Because of the limited
shoreline of one super-continental mass, Triassic marine deposits are globally
relatively rare; despite their prominence in Western Europe, where the Triassic
was first studied. In North America, for example, marine deposits are limited
to a few exposures in the west. Thus Triassic stratigraphy is mostly based on
organisms living in lagoons and hypersaline environments, such as *Estheria*
crustaceans and terrestrial vertebrates.

Jurassic Period

The **Jurassic** Period extends from about 201.3 ± 0.2 to 145.0 Ma. During the
early Jurassic, the supercontinent Pangaea broke up into the northern supercontinent Laurasia and the southern supercontinent Gondwana; the Gulf of
Mexico opened in the new rift between North America and what is now Mexico's Yucatan Peninsula. The Jurassic North Atlantic Ocean was relatively
narrow, while the South Atlantic did not open until the following Cretaceous
Period, when Gondwana itself rifted apart. The Tethys Sea closed, and the

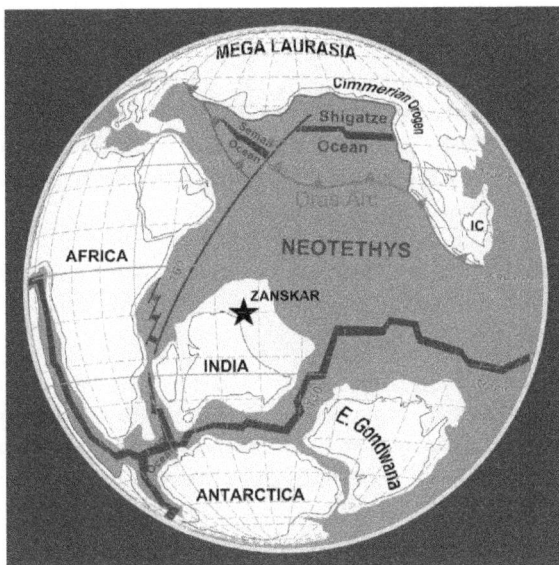

Figure 6: *Plate tectonics- 100 Ma, Cretaceous period*

Neotethys basin appeared. Climates were warm, with no evidence of glaciation. As in the Triassic, there was apparently no land near either pole, and no extensive ice caps existed. The Jurassic geological record is good in western Europe, where extensive marine sequences indicate a time when much of the continent was submerged under shallow tropical seas; famous locales include the Jurassic Coast World Heritage Site and the renowned late Jurassic *lagerstätten* of Holzmaden and Solnhofen. In contrast, the North American Jurassic record is the poorest of the Mesozoic, with few outcrops at the surface. Though the epicontinental Sundance Sea left marine deposits in parts of the northern plains of the United States and Canada during the late Jurassic, most exposed sediments from this period are continental, such as the alluvial deposits of the Morrison Formation. The first of several massive batholiths were emplaced in the northern Cordillera beginning in the mid-Jurassic, marking the Nevadan orogeny.[16] Important Jurassic exposures are also found in Russia, India, South America, Japan, Australasia and the United Kingdom.

Cretaceous Period

The **Cretaceous** Period extends from circa 145[17] million years ago to 66[18] million years ago.

During the Cretaceous, the late Paleozoic-early Mesozoic supercontinent of Pangaea completed its breakup into present day continents, although their positions were substantially different at the time. As the Atlantic Ocean widened, the convergent-margin orogenies that had begun during the Jurassic continued in the North American Cordillera, as the Nevadan orogeny was followed by the Sevier and Laramide orogenies. Though Gondwana was still intact in the beginning of the Cretaceous, Gondwana itself broke up as South America, Antarctica and Australia rifted away from Africa (though India and Madagascar remained attached to each other); thus, the South Atlantic and Indian Oceans were newly formed. Such active rifting lifted great undersea mountain chains along the welts, raising eustatic sea levels worldwide.

To the north of Africa the Tethys Sea continued to narrow. Broad shallow seas advanced across central North America (the Western Interior Seaway) and Europe, then receded late in the period, leaving thick marine deposits sandwiched between coal beds. At the peak of the Cretaceous transgression, one-third of Earth's present land area was submerged.[19] The Cretaceous is justly famous for its chalk; indeed, more chalk formed in the Cretaceous than in any other period in the Phanerozoic. Mid-ocean ridge activity—or rather, the circulation of seawater through the enlarged ridges—enriched the oceans in calcium; this made the oceans more saturated, as well as increased the bioavailability of the element for calcareous nanoplankton. These widespread carbonates and other sedimentary deposits make the Cretaceous rock record especially fine. Famous formations from North America include the rich marine fossils of Kansas's Smoky Hill Chalk Member and the terrestrial fauna of the late Cretaceous Hell Creek Formation. Other important Cretaceous exposures occur in Europe and China. In the area that is now India, massive lava beds called the Deccan Traps were laid down in the very late Cretaceous and early Paleocene.

Cenozoic Era

The **Cenozoic** Era covers the 66 million years since the Cretaceous–Paleogene extinction event up to and including the present day. By the end of the Mesozoic era, the continents had rifted into nearly their present form. Laurasia became North America and Eurasia, while Gondwana split into South America, Africa, Australia, Antarctica and the Indian subcontinent, which collided with the Asian plate. This impact gave rise to the Himalayas. The Tethys Sea, which had separated the northern continents from Africa and India, began to close up, forming the Mediterranean sea.

Paleogene Period

The **Paleogene** (alternatively **Palaeogene**) Period is a unit of geologic time that began 66 and ended 23.03 Ma and comprises the first part of the Cenozoic Era. This period consists of the Paleocene, Eocene and Oligocene Epochs.

Paleocene Epoch

The **Paleocene**, lasted from 66^{18} million years ago to 56^{20} million years ago.

In many ways, the Paleocene continued processes that had begun during the late Cretaceous Period. During the Paleocene, the continents continued to drift toward their present positions. Supercontinent Laurasia had not yet separated into three continents. Europe and Greenland were still connected. North America and Asia were still intermittently joined by a land bridge, while Greenland and North America were beginning to separate.[21] The Laramide orogeny of the late Cretaceous continued to uplift the Rocky Mountains in the American west, which ended in the succeeding epoch. South and North America remained separated by equatorial seas (they joined during the Neogene); the components of the former southern supercontinent Gondwana continued to split apart, with Africa, South America, Antarctica and Australia pulling away from each other. Africa was heading north toward Europe, slowly closing the Tethys Ocean, and India began its migration to Asia that would lead to a tectonic collision and the formation of the Himalayas.

Eocene Epoch

During the **Eocene** (56^{20} million years ago - 33.9^{22} million years ago), the continents continued to drift toward their present positions. At the beginning of the period, Australia and Antarctica remained connected, and warm equatorial currents mixed with colder Antarctic waters, distributing the heat around the world and keeping global temperatures high. But when Australia split from the southern continent around 45 Ma, the warm equatorial currents were deflected away from Antarctica, and an isolated cold water channel developed between the two continents. The Antarctic region cooled down, and the ocean surrounding Antarctica began to freeze, sending cold water and ice floes north, reinforcing the cooling. The present pattern of ice ages began about 40^4 million years ago.Wikipedia:Citation needed

The northern supercontinent of Laurasia began to break up, as Europe, Greenland and North America drifted apart. In western North America, mountain building started in the Eocene, and huge lakes formed in the high flat basins among uplifts. In Europe, the Tethys Sea finally vanished, while the uplift of the Alps isolated its final remnant, the Mediterranean, and created another shallow sea with island archipelagos to the north. Though the North

Atlantic was opening, a land connection appears to have remained between North America and Europe since the faunas of the two regions are very similar. India continued its journey away from Africa and began its collision with Asia, creating the Himalayan orogeny.

Oligocene Epoch

The **Oligocene** Epoch extends from about 34^{23} million years ago to 23^{24} million years ago. During the Oligocene the continents continued to drift toward their present positions.

Antarctica continued to become more isolated and finally developed a permanent ice cap. Mountain building in western North America continued, and the Alps started to rise in Europe as the African plate continued to push north into the Eurasian plate, isolating the remnants of Tethys Sea. A brief marine incursion marks the early Oligocene in Europe. There appears to have been a land bridge in the early Oligocene between North America and Europe since the faunas of the two regions are very similar. During the Oligocene, South America was finally detached from Antarctica and drifted north toward North America. It also allowed the Antarctic Circumpolar Current to flow, rapidly cooling the continent.

Neogene Period

The **Neogene** Period is a unit of geologic time starting 23.03 Ma. and ends at 2.588 Mya. The Neogene Period follows the Paleogene Period. The Neogene consists of the Miocene and Pliocene and is followed by the Quaternary Period.

Miocene Epoch

The **Miocene** extends from about 23.03 to 5.333 Ma.

During the Miocene continents continued to drift toward their present positions. Of the modern geologic features, only the land bridge between South America and North America was absent, the subduction zone along the Pacific Ocean margin of South America caused the rise of the Andes and the southward extension of the Meso-American peninsula. India continued to collide with Asia. The Tethys Seaway continued to shrink and then disappeared as Africa collided with Eurasia in the Turkish-Arabian region between 19 and 12 Ma (ICS 2004). Subsequent uplift of mountains in the western Mediterranean region and a global fall in sea levels combined to cause a temporary drying up of the Mediterranean Sea resulting in the Messinian salinity crisis near the end of the Miocene.

Pliocene Epoch

The **Pliocene** extends from 5.333^{25} million years ago to 2.588^{26} million years ago. During the Pliocene continents continued to drift toward their present positions, moving from positions possibly as far as 250 kilometres (155 mi) from their present locations to positions only 70 km from their current locations.

South America became linked to North America through the Isthmus of Panama during the Pliocene, bringing a nearly complete end to South America's distinctive marsupial faunas. The formation of the Isthmus had major consequences on global temperatures, since warm equatorial ocean currents were cut off and an Atlantic cooling cycle began, with cold Arctic and Antarctic waters dropping temperatures in the now-isolated Atlantic Ocean. Africa's collision with Europe formed the Mediterranean Sea, cutting off the remnants of the Tethys Ocean. Sea level changes exposed the land-bridge between Alaska and Asia. Near the end of the Pliocene, about 2.58^{27} million years ago (the start of the Quaternary Period), the current ice age began. The polar regions have since undergone repeated cycles of glaciation and thaw, repeating every 40,000–100,000 years.

Quaternary Period

Pleistocene Epoch

The **Pleistocene** extends from 2.588^{26} million years ago to 11,700 years before present. The modern continents were essentially at their present positions during the Pleistocene, the plates upon which they sit probably having moved no more than 100 kilometres (62 mi) relative to each other since the beginning of the period.

Holocene Epoch

The **Holocene** Epoch began approximately 11,700 calendar years before present and continues to the present. During the Holocene, continental motions have been less than a kilometer.

The last glacial period of the current ice age ended about 10,000 years ago. Ice melt caused world sea levels to rise about 35 metres (115 ft) in the early part of the Holocene. In addition, many areas above about 40 degrees north latitude had been depressed by the weight of the Pleistocene glaciers and rose as much as 180 metres (591 ft) over the late Pleistocene and Holocene, and are still rising today. The sea level rise and temporary land depression allowed temporary marine incursions into areas that are now far from the sea. Holocene marine fossils are known from Vermont, Quebec, Ontario and Michigan. Other than higher latitude temporary marine incursions associated with glacial depression, Holocene fossils are found primarily in lakebed, floodplain and cave

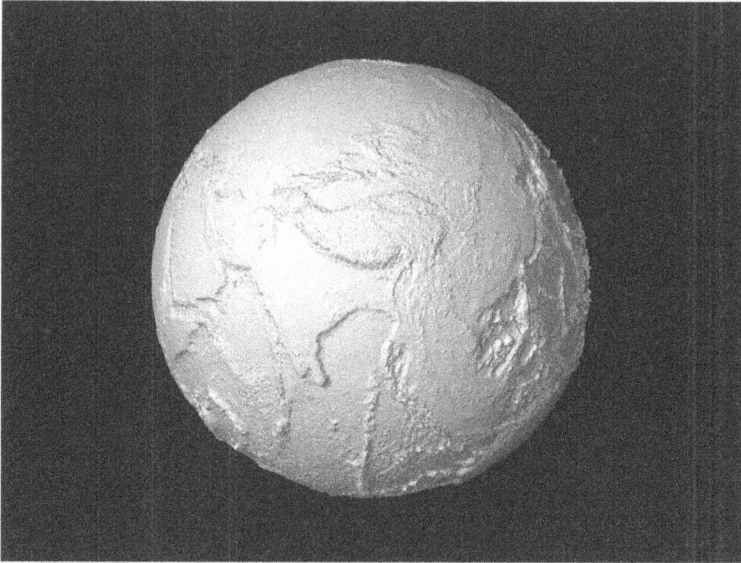

Figure 7: *Current Earth - without water (click/enlarge to "spin" 3D-globe).*

deposits. Holocene marine deposits along low-latitude coastlines are rare because the rise in sea levels during the period exceeds any likely upthrusting of
non-glacial origin. Post-glacial rebound in Scandinavia resulted in the emergence of coastal areas around the Baltic Sea, including much of Finland. The
region continues to rise, still causing weak earthquakes across Northern Europe. The equivalent event in North America was the rebound of Hudson
Bay, as it shrank from its larger, immediate post-glacial Tyrrell Sea phase, to
near its present boundaries.

Further reading

<templatestyles src="Template:Refbegin/styles.css" />

- Stanley, Steven M. (1999). *Earth system history* (New ed.). New York:
  W. H. Freeman. ISBN 978-0-7167-3377-5.

External links

- Cosmic Evolution[28] — a detailed look at events from the origin of the
  universe to the present

- Valley, John W. " A Cool Early Earth?[29]" *Scientific American.* 2005 Oct:58–65. – discusses the timing of the formation of the oceans and other major events in Eh's early history.
- Davies, Paul. " Quantum leap of life[30]". *The Guardian.* 2005 Dec 20. – discusses speculation into the role of quantum systems in the origin of life
- Evolution timeline[31] (uses Shockwave). Animated story of life since about 13,700,000,000 shows everything from the big bang to the formation of the earth and the development of bacteria and other organisms to the ascent of man.
- Theory of the Earth & Abstract of the Theory of the Earth[32]
- Paleomaps Since 600 Ma (Mollweide Projection, Longitude 0)[33]
- Paleomaps Since 600 Ma (Mollweide Projection, Longitude 180)[34]
- Ageing the Earth[35] on *In Our Time* at the BBC

Introduction

Precambrian

Precambrian Eon
4600–541 million years ago

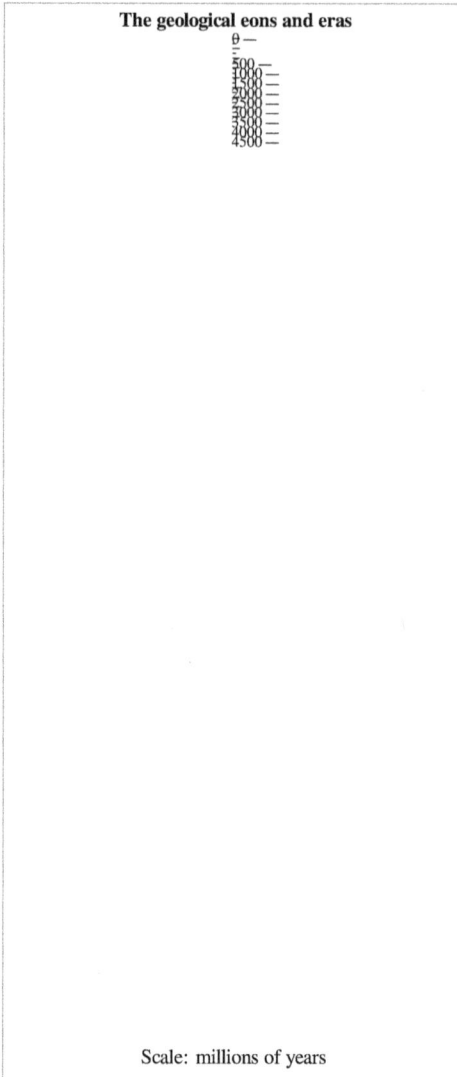

The geological eons and eras

Scale: millions of years

The **Precambrian** (or **Pre-Cambrian**, sometimes abbreviated **p€**, or **Crypto-zoic**) is the earliest part of Earth's history, set before the current Phanerozoic Eon. The Precambrian is so named because it preceded the Cambrian, the first period of the Phanerozoic eon, which is named after Cambria, the Latinised name for Wales, where rocks from this age were first studied. The Precambrian accounts for 88% of the Earth's geologic time.

The Precambrian (colored green in the timeline figure) is an informal unit of geologic time, subdivided into three eons (Hadean, Archean, Proterozoic) of the geologic time scale. It spans from the formation of Earth about 4.6 billion years ago (Ga) to the beginning of the Cambrian Period, about 541 million years ago (Ma), when hard-shelled creatures first appeared in abundance.

Overview

Relatively little is known about the Precambrian, despite it making up roughly seven-eighths of the Earth's history, and what is known has largely been discovered from the 1960s onwards. The Precambrian fossil record is poorer than that of the succeeding Phanerozoic, and fossils from the Precambrian (e.g. stromatolites) are of limited biostratigraphic use. This is because many Precambrian rocks have been heavily metamorphosed, obscuring their origins, while others have been destroyed by erosion, or remain deeply buried beneath Phanerozoic strata.

It is thought that the Earth coalesced from material in orbit around the Sun at roughly 4,543 Ma, and may have been struck by a very large (Mars-sized) planetesimal shortly after it formed, splitting off material that formed the Moon (see Giant impact hypothesis). A stable crust was apparently in place by 4,433 Ma, since zircon crystals from Western Australia have been dated at $4,404 \pm 8$ Ma.

The term "Precambrian" is recognized by the International Commission on Stratigraphy as the only "supereon" in geologic time; Wikipedia:Citation needed it is so called because it includes the Hadean ($\sim$4.6—4 billion), Archean (4—2.5 billion), and Proterozoic (2.5 billion—541 million) eons. (There is only one other eon: the Phanerozoic, 541 million-present.) "Precambrian" is still used by geologists and paleontologists for general discussions not requiring the more specific eon names. As of 2010[36], the United States Geological Survey considers the term informal, lacking a stratigraphic rank.

Life forms

A specific date for the origin of life has not been determined. Carbon found in 3.8 billion-year-old rocks (Archean eon) from islands off western Greenland may be of organic origin. Well-preserved microscopic fossils of bacteria older than 3.46 billion years have been found in Western Australia. Probable fossils 100 million years older have been found in the same area. However, there is evidence that live could have evolved over 4.280 billion years ago. There is a

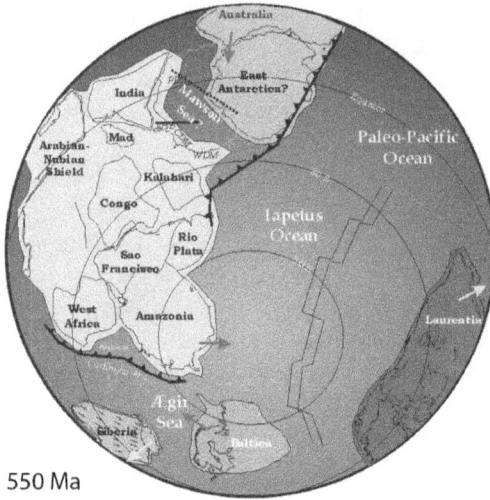

550 Ma

Figure 8: *Landmass positions near the end of the Precambrian*

fairly solid record of bacterial life throughout the remainder (Proterozoic eon) of the Precambrian.

Excluding a few contested reports of much older forms from North America and India, the first complex multicellular life forms seem to have appeared at roughly 1500 Ma, in the Mesoproterozoic era of the Proterozoic eon.Wikipedia:Citation needed Fossil evidence from the later Ediacaran period of such complex life comes from the Lantian formation, at least 580 million years ago. A very diverse collection of soft-bodied forms is found in a variety of locations worldwide and date to between 635 and 542 Ma. These are referred to as Ediacaran or Vendian biota. Hard-shelled creatures appeared toward the end of that time span, marking the beginning of the Phanerozoic eon. By the middle of the following Cambrian period, a very diverse fauna is recorded in the Burgess Shale, including some which may represent stem groups of modern taxa. The increase in diversity of lifeforms during the early Cambrian is called the Cambrian explosion of life.

While land seems to have been devoid of plants and animals, cyanobacteria and other microbes formed prokaryotic mats that covered terrestrial areas.

Tracks from an animal with leg like appendages have been found in what was mud 551 million years ago.[37]

Figure 9: *Weathered Precambrian pillow lava in the Temagami Greenstone Belt of the Canadian Shield*

Planetary environment and the oxygen catastrophe

Evidence of the details of plate motions and other tectonic activity in the Precambrian has been poorly preserved. It is generally believed that small proto-continents existed prior to 4280 Ma, and that most of the Earth's landmasses collected into a single supercontinent around 1130 Ma. The supercontinent, known as Rodinia, broke up around 750 Ma. A number of glacial periods have been identified going as far back as the Huronian epoch, roughly 2400–2100 Ma. One of the best studied is the Sturtian-Varangian glaciation, around 850–635 Ma, which may have brought glacial conditions all the way to the equator, resulting in a "Snowball Earth".

The atmosphere of the early Earth is not well understood. Most geologists believe it was composed primarily of nitrogen, carbon dioxide, and other relatively inert gases, and was lacking in free oxygen. There is, however, evidence that an oxygen-rich atmosphere existed since the early Archean.

At present, it is still believed that molecular oxygen was not a significant fraction of Earth's atmosphere until after photosynthetic life forms evolved and began to produce it in large quantities as a byproduct of their metabolism. This radical shift from a chemically inert to an oxidizing atmosphere caused an ecological crisis, sometimes called the oxygen catastrophe. At first, oxygen would have quickly combined with other elements in Earth's crust, primarily iron, removing it from the atmosphere. After the supply of oxidizable surfaces ran out, oxygen would have begun to accumulate in the atmosphere, and the modern high-oxygen atmosphere would have developed. Evidence for this lies in older rocks that contain massive banded iron formations that were laid down as iron oxides.

Subdivisions

Life timeline

0 —
500
1000
1500
2000
2500
3000
3500
4000
4500

Axis scale: million years

🖐

Also see: *Human timeline* and *Nature timeline*

A terminology has evolved covering the early years of the Earth's existence, as radiometric dating has allowed real dates to be assigned to specific formations and features.[38] The Precambrian is divided into three eons: the Hadean (4600–4000 Ma), Archean (4000-2500 Ma) and Proterozoic (2500-541 Ma). See Timetable of the Precambrian.

- Proterozoic: this eon refers to the time from the lower Cambrian boundary, 541 Ma, back through 2500 Ma. As originally used, it was a synonym for "Precambrian" and hence included everything prior to the Cambrian boundary. The Proterozoic eon is divided into three eras: the Neoproterozoic, Mesoproterozoic and Paleoproterozoic.
  - Neoproterozoic: The youngest geologic era of the Proterozoic Eon, from the Cambrian Period lower boundary (541 Ma) back to 1000 Ma. The Neoproterozoic corresponds to Precambrian Z rocks of older North American geology.
    - Ediacaran: The youngest geologic period within the Neoproterozoic Era. The "2012 Geologic Time Scale" dates it from 541 to 635 Ma. In this period the Ediacaran fauna appeared.
    - Cryogenian: The middle period in the Neoproterozoic Era: 635-720 Ma.
    - Tonian: the earliest period of the Neoproterozoic Era: 720-1000 Ma.
  - Mesoproterozoic: the middle era of the Proterozoic Eon, 1000-1600 Ma. Corresponds to "Precambrian Y" rocks of older North American geology.
  - Paleoproterozoic: oldest era of the Proterozoic Eon, 1600-2500 Ma. Corresponds to "Precambrian X" rocks of older North American geology.
- Archean Eon: 2500-4000 Ma.
- Hadean Eon: 4000–4600 Ma. This term was intended originally to cover the time before any preserved rocks were deposited, although some zircon crystals from about 4400 Ma demonstrate the existence of crust in the Hadean Eon. Other records from Hadean time come from the moon and meteorites.

It has been proposed that the Precambrian should be divided into eons and eras that reflect stages of planetary evolution, rather than the current scheme based upon numerical ages. Such a system could rely on events in the stratigraphic record and be demarcated by GSSPs. The Precambrian could be divided into five "natural" eons, characterized as follows:[39]

1. Accretion and differentiation: a period of planetary formation until giant Moon-forming impact event.
2. Hadean: dominated by heavy bombardment from about 4.51 Ga (possibly including a Cool Early Earth period) to the end of the Late Heavy Bombardment period.
3. Archean: a period defined by the first crustal formations (the Isua greenstone belt) until the deposition of banded iron formations due to increasing atmospheric oxygen content.
4. Transition: a period of continued iron banded formation until the first continental red beds.
5. Proterozoic: a period of modern plate tectonics until the first animals.

Precambrian supercontinents

The movement of Earth's plates has caused the formation and break-up of continents over time, including occasional formation of a supercontinent containing most or all of the landmass. The earliest known supercontinent was Vaalbara. It formed from proto-continents and was a supercontinent 3.636 billion years ago. Vaalbara broke up c. 2.845–2.803 Ga ago. The supercontinent Kenorland was formed c. 2.72 Ga ago and then broke sometime after 2.45–2.1 Ga into the proto-continent cratons called Laurentia, Baltica, Yilgarn craton, and Kalahari. The supercontinent Columbia or Nuna formed 2.06–1.82 billion years ago and broke up about 1.5–1.35 billion years ago.Wikipedia:Verifiability The supercontinent Rodinia is thought to have formed about 1.13–1.071 billion years ago, to have embodied most or all of Earth's continents and to have broken up into eight continents around 750–600 million years ago.

Further reading

- Valley, John W., William H. Peck, Elizabeth M. King (1999) *Zircons Are Forever*, The Outcrop for 1999, University of Wisconsin-Madison Wgeology.wisc.edu[40] – *Evidence from detrital zircons for the existence of continental crust and oceans on the Earth 4.4 Gyr ago* Accessed Jan. 10, 2006
- Wilde, S. A.; Valley, J. W.; Peck, W. H.; Graham, C. M. (2001). "Evidence from detrital zircons for the existence of continental crust and oceans on the Earth 4.4 Gyr ago". *Nature*. **409** (6817): 175–178. doi: 10.1038/35051550[41]. PMID 11196637[42].
- Wyche, S.; Nelson, D. R.; Riganti, A. (2004). "4350–3130 Ma detrital zircons in the Southern Cross Granite–Greenstone Terrane, Western Australia: implications for the early evolution of the Yilgarn Craton".

Australian Journal of Earth Sciences. **51** (1): 31–45. Bibcode: 2004Au-
JES..51...31W[43]. doi: 10.1046/j.1400-0952.2003.01042.x[44].

Wikimedia Commons has media related to ***Precambrian***.

External links

- Late Precambrian Supercontinent and Ice House World[45] from the Paleomap Project

Hadean

Hadean Eon
4600–4000 million years ago

The geological eons and eras

```
θ —
 —
500 —
1000 —
1500 —
2000 —
2500 —
3000 —
3500 —
4000 —
4500 —
```

Scale: millions of years

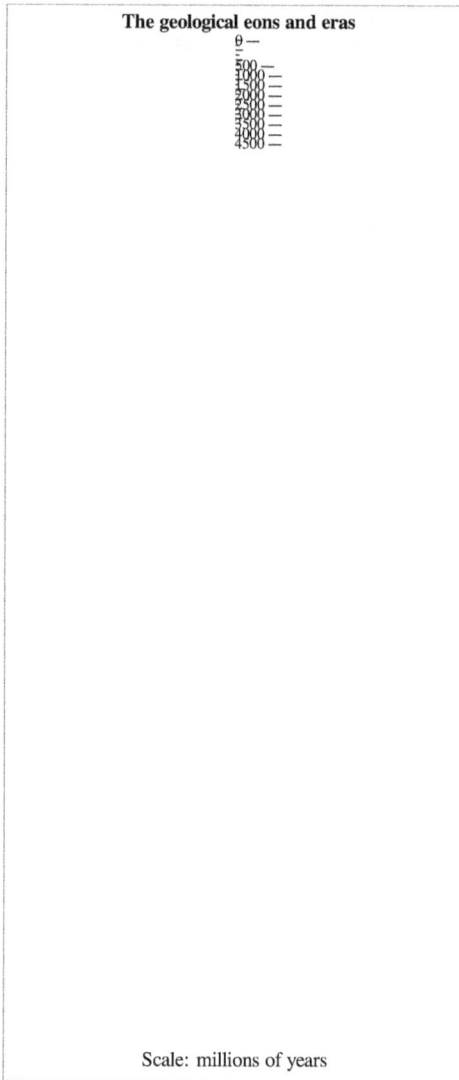

The **Hadean** (/ˈheɪdiən/) is a geologic eon of the Earth predating the Archean. It began with the formation of the Earth about 4.6 billion years ago and ended, as defined by the ICS, 4 billion years ago. As of 2016[46], the ICS describes its status as *"informal"*. Geologist Preston Cloud coined the term in 1972, originally to label the period before the earliest-known rocks on Earth. W. Brian Harland later coined an almost synonymous term: the **"Priscoan period"**. Other, older texts simply refer to the eon as the **Pre-Archean**.

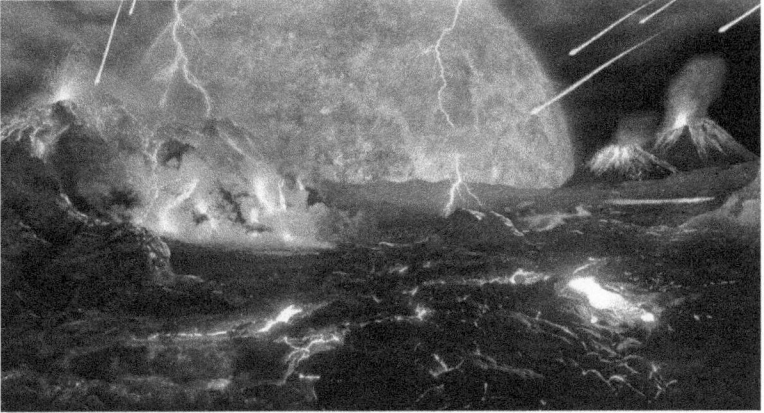

Figure 10: *Artist's impression of a Hadean landscape*

Etymology

Life timeline

Axis scale: million years

🖑

Also see: *Human timeline* and *Nature timeline*

"Hadean" (from Hades, the Greek god of the underworld) describes the hellish conditions then prevailing on Earth: the planet had just formed and was still very hot owing to its recent accretion, the abundance of short-lived radioactive elements, and frequent collisions with other Solar System bodies.

Subdivisions

Since few geological traces of this eon remain on Earth, there is no official subdivision. However, the Lunar geologic timescale embraces several major divisions relating to the Hadean, so these are sometimes used in an informal sense to refer to the same periods of time on Earth.

The Lunar divisions are:

- Pre-Nectarian, from the formation of the Moon's crust (4,533[47] million years ago) up to about 3,920[48] million years ago
- Nectarian ranging from 3,920[48] million years ago up to about 3,850[49] million years ago, in a time when the Late Heavy Bombardment, according to that theory, was in a stage of decline.

In 2010, an alternative scale was proposed that includes the addition of the Chaotian and Prenephelean Eons preceding the Hadean, and divides the Hadean into three eras with two periods each. The Paleohadean era consists of the Hephaestean (4.5–4.4 Ga) and the Jacobian periods (4.4–4.3 Ga). The Mesohadean is divided into the Canadian (4.3–4.2 Ga) and the Procrustean periods (4.2–4.1 Ga). The Neohadean is divided into the Acastan (4.1–4.0 Ga) and the Promethean periods (4.0–3.9 Ga). As of February 2017[46], this has not been adopted by the IUGS.

Hadean rocks

In the last decades of the 20th century geologists identified a few Hadean rocks from Western Greenland, Northwestern Canada, and Western Australia. In 2015, traces of carbon minerals interpreted as "remains of biotic life" were found in 4.1-billion-year-old rocks in Western Australia.[50]

The oldest dated zircon crystals, enclosed in a metamorphosed sandstone conglomerate in the Jack Hills of the Narryer Gneiss Terrane of Western Australia, date to 4.404 ± 0.008 Ga. This zircon is a slight outlier, with the oldest consistently-dated zircon falling closer to 4.35 Ga—around 200 million years after the hypothesized time of the Earth's formation.

In many other areas, xenocryst (or relict) Hadean zircons enclosed in older rocks indicate that younger rocks have formed on older terranes and have incorporated some of the older material. One example occurs in the Guiana shield from the Iwokrama Formation of southern Guyana where zircon cores have been dated at 4.22 Ga.

Atmosphere and oceans

A sizeable quantity of water would have been in the material that formed the Earth. Water molecules would have escaped Earth's gravity more easily when it was less massive during its formation. Hydrogen and helium are expected to continually escape (even to the present day) due to atmospheric escape. Part of the ancient planet is theorized to have been disrupted by the impact that created the Moon, which should have caused melting of one or two large regions of the Earth. Earth's present composition suggests that there was not complete remelting as it is difficult to completely melt and mix huge rock masses.[51] However, a fair fraction of material should have been vaporized by this impact, creating a *rock vapor atmosphere* around the young planet. The rock vapor would have condensed within two thousand years, leaving behind hot volatiles which probably resulted in a heavy CO_2 atmosphere with hydrogen and water vapor. Liquid water oceans existed despite the surface temperature of 230 °C (446 °F) because at an atmospheric pressure of above 27 atmospheres, caused by the heavy CO_2 atmosphere, water is still liquid. As cooling continued, subduction and dissolving in ocean water removed most CO_2 from the atmosphere but levels oscillated wildly as new surface and mantle cycles appeared.

Studies of zircons have found that liquid water must have existed as long ago as 4.4 billion years ago, very soon after the formation of the Earth.[52,53,54] This requires the presence of an atmosphere. The cool early Earth theory covers a range from about 4.4 to about 4.1 billion years.

A September 2008 study of zircons found that Australian Hadean rock holds minerals pointing to the existence of plate tectonics as early as 4 billion years. If this is true, the time when Earth finished its transition from having a hot, molten surface and atmosphere full of carbon dioxide, to being very much like it is today, can be roughly dated to about 4.0 billion years ago. The actions of plate tectonics and the oceans trapped vast amounts of carbon dioxide, thereby eliminating the greenhouse effect and leading to a much cooler surface temperature and the formation of solid rock, and possibly even life.

Further reading

- Hopkins, Michelle; Harrison, T. Mark; Manning, Craig E. (2008), "Low heat flow inferred from >4 Gyr zircons suggests Hadean plate boundary interactions", *Nature*, **456** (7221): 493–496, Bibcode: 2008Natur.456..493H[55], doi: 10.1038/nature07465[56], PMID 19037314[57].
- Valley, John W.; Peck, William H.; King, Elizabeth M. (1999), "Zircons Are Forever"[58], *The Outcrop for 1999, University of Wisconsin-Madison*, retrieved January 10, 2006 – *Evidence from detrital zircons for the existence of continental crust and oceans on the Earth 4.4 Gyr ago.*
- Wilde, S. A.; Valley, J. W.; Peck, W. H. & Graham, C. M. (2001), "Evidence from detrital zircons for the existence of continental crust and oceans on the Earth 4.4 Gyr ago", *Nature*, **409** (6817): 175–178, doi: 10.1038/35051550[59], PMID 11196637[60].
- Wyche, S.; Nelson, D. R. & Riganti, A. (2004), "4350–3130 Ma detrital zircons in the Southern Cross Granite–Greenstone Terrane, Western Australia: implications for the early evolution of the Yilgarn Craton", *Australian Journal of Earth Sciences*, **51** (1): 31–45, Bibcode: 2004AuJES..51...31W[61], doi: 10.1046/j.1400-0952.2003.01042.x[62].
- Carley, Tamara L.; et al. (2014), "Iceland is not a magmatic analog for the Hadean: Evidence from the zircon record", *Earth and Planetary Science Letters*, **405** (1): 85–97, Bibcode: 2014E&PSL.405...85C[63], doi: 10.1016/j.epsl.2014.08.015[64].

External links

Wikimedia Commons has media related to *Hadean*.

- Palaeos.org: Hadean eon[65]
- Peripatus.nz: Description of the Hadean Era[66]
- Astronoo.com: Hell of the Hadean[67]

Archean

Archean Eon
4000–2500 million years ago

The geological eons and eras

0 —

500
1000
1500
2000
2500
3000
3500
4000
4500

Scale: millions of years

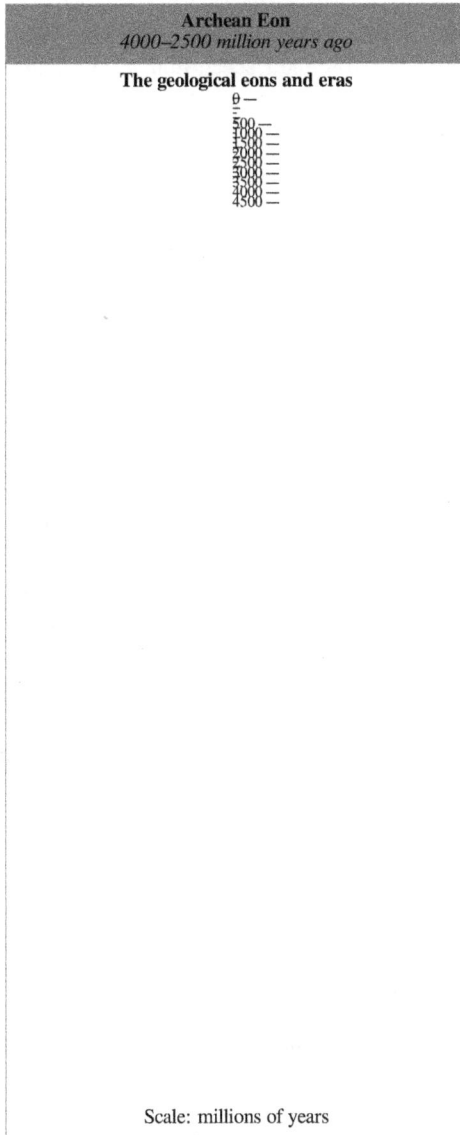

The **Archean** Eon (/ɑːrˈkiːən/, also spelled **Archaean** or **Archæan**) is one of the four geologic eons of Earth history, occurring 4,000 to 2,500[68] million years ago (4 to 2.5 billion years ago). During the Archean, the Earth's crust had cooled enough to allow the formation of continents and life started to form.

Etymology and changes in classification

Archean (or Archaean) comes from the ancient Greek Αρχή (*Arkhē*), meaning "beginning, origin". Its earliest use is from 1872, when it meant "of the earliest geological age."[69] Before the Hadean Eon was recognized, the Archean spanned Earth's early history from its creation about 4,540 million years ago until 2,500 million years ago.Wikipedia:Citation needed

Instead of being based on stratigraphy, the beginning and end of the Archean Eon are defined chronometrically. The eon's lower boundary or starting point of 4 Gya (4 billion years ago) is officially recognized by the International Commission on Stratigraphy.

Geology

When the Archean began, the Earth's heat flow was nearly three times as high as it is today, and it was still twice the current level at the transition from the Archean to the Proterozoic (2,500 million years ago). The extra heat was the result of a mix of remnant heat from planetary accretion, from the formation of the Earth's core, and from the disintegration of radioactive elements.

Although a few mineral grains are known to be Hadean, the oldest rock formations exposed on the surface of the Earth are Archean. Archean rocks are found in Greenland, Siberia, the Canadian Shield, Montana and Wyoming (exposed parts of the Wyoming Craton), the Baltic Shield, Scotland, India, Brazil, western Australia, and southern Africa. Granitic rocks predominate throughout the crystalline remnants of the surviving Archean crust. Examples include great melt sheets and voluminous plutonic masses of granite, diorite, layered intrusions, anorthosites and monzonites known as sanukitoids. Archean Eon rocks are often heavily metamorphized deep-water sediments, such as graywackes, mudstones, volcanic sediments, and banded iron formations. Volcanic activity was considerably higher than today, with numerous lava eruptions, including unusual types such as komatiite.Wikipedia:Citation needed Carbonate rocks are rare, indicating that the oceans were more acidic due to dissolved carbon dioxide than during the Proterozoic. Greenstone belts are typical Archean formations, consisting of alternating units of metamorphosed mafic igneous and sedimentary rocks. The metamorphosed igneous rocks were derived from volcanic island arcs, while the metamorphosed sediments represent deep-sea sediments eroded from the neighboring island arcs and deposited in a forearc basin. Greenstone belts, being both types of metamorphosed rock, represent sutures between the protocontinents.:302–03

The Earth's continents started to form in the Archean, although details about their formation are still being debated, due to lack of extensive geological evidence. One hypothesis is that rocks that are now in India, western Australia,

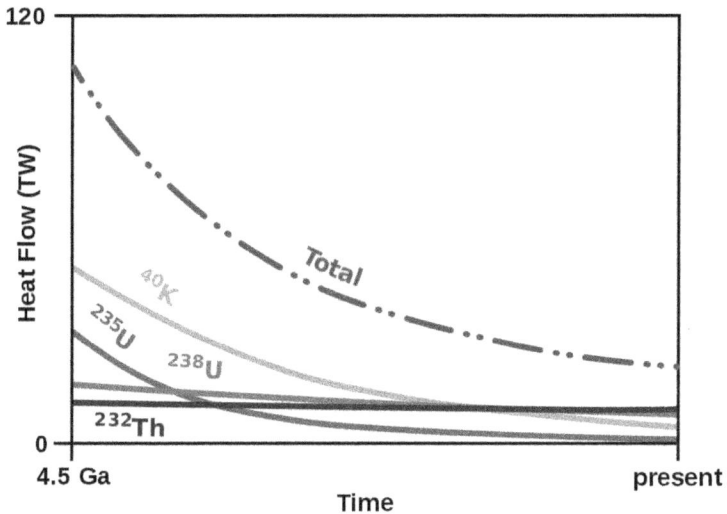

Figure 11: *The evolution of Earth's radiogenic heat flow over time*

and southern Africa formed a continent called Ur as of 3,100 Ma. A differing conflicting hypothesis is that rocks from western Australia and southern Africa were assembled in a continent called Vaalbara as far back as 3,600 Ma. Although the first continents formed during this eon, rock of this age makes up only 7% of the present world's cratons; even allowing for erosion and destruction of past formations, evidence suggests that only 5–40% of the present area of continents formed during the Archean.:301–02

By the end of the Archaean c. 2500 Ma, plate tectonic activity may have been similar to that of the modern Earth. There are well-preserved sedimentary basins, and evidence of volcanic arcs, intracontinental rifts, continent-continent collisions and widespread globe-spanning orogenic events suggesting the assembly and destruction of one and perhaps several supercontinents. Liquid water was prevalent, and deep oceanic basins are known to have existed attested by the presence of banded iron formations, chert beds, chemical sediments and pillow basalts.

Environment

The Archean atmosphere is thought to have nearly lacked free oxygen. As-
tronomers think that the Sun had about 70–75 percent of the present luminos-
ity, yet temperatures on Earth appear to have been near modern levels after
only 500 Ma of Earth's formation (the faint young Sun paradox). The presence of liquid water is evidenced by certain highly deformed gneisses produced
by metamorphism of sedimentary protoliths. The moderate temperatures may
reflect the presence of greater amounts of greenhouse gases than later in the
Earth's history. Alternatively, Earth's albedo may have been lower at the time,
due to less land area and cloud cover.

Early life

Life timeline

```
0 —
500 —
1000 —
1500 —
2000 —
2500 —
3000 —
3500 —
4000 —
4500 —
```

Axis scale: million years

🖑

Also see: *Human timeline* and *Nature timeline*

The processes that gave rise to life on Earth are not completely understood, but there is substantial evidence that life came into existence either near the end of the Hadean Eon or early in the Archean Eon.

The earliest evidence for life on Earth are graphite of biogenic origin found in 3.7-billion-year-old metasedimentary rocks discovered in Western Greenland.

The earliest identifiable fossils consist of stromatolites, which are microbial mats formed in shallow water by cyanobacteria. The earliest stromatolites are found in 3.48 billion-year-old sandstone discovered in Western Australia. Stromatolites are found throughout the Archean and become common late in the Archean.:[307] Cyanobacteria were instrumental in creating free oxygen in the atmosphere.

Further evidence for early life is found in 3.47-billon-year-old baryte, in the Warrawoona Group of Western Australia. This mineral shows sulfur fractionation of as much as 21.1%, which is evidence of sulfate-reducing bacteria that metabolize sulfur-32 more readily than sulfur-34.

Evidence of life in the Late Hadean is more controversial. In 2015, biogenic carbon has been detected in zircons dated to 4.1 billion years ago, but this evidence is preliminary and needs validation.[70]

Earth was very hostile to life before 4.2–4.3 Ga and the conclusion is that before the Archean Eon, life as we know it would have been challenged by these environmental conditions. While life could have arisen before the Archean, the conditions necessary to sustain life could not have occurred until the Archean Eon.

Life in the Archean was limited to simple single-celled organisms (lacking nuclei), called Prokaryota. In addition to the domain Bacteria, microfossils of the domain Archaea have also been identified. There are no known eukaryotic fossils from the earliest Archean, though they might have evolved during the Archean without leaving any.:[306,323] No fossil evidence has been discovered for ultramicroscopic intracellular replicators such as viruses.

External links

Wikimedia Commons has media related to *Archean*.

Wikisourcehas the text of the 1911 *Encyclopædia Britannica*article *Archean System*.

- GeoWhen Database[71]
- When Did Plate Tectonics Begin?[72]

Proterozoic

Proterozoic Eon
2500–541 million years ago

The geological eons and eras

θ —
500 —
1000 —
1500 —
2000 —
2500 —
3000 —
3500 —
4000 —
4500 —

Scale: millions of years

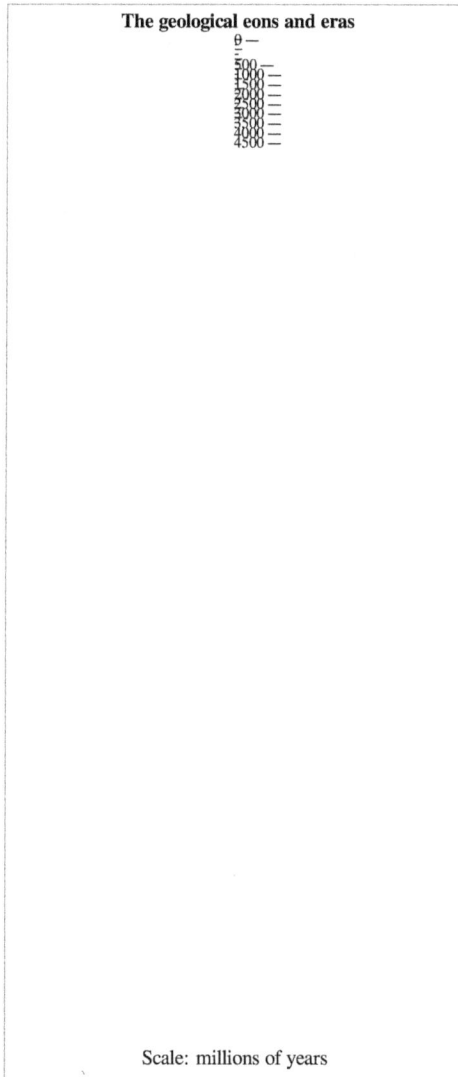

The **Proterozoic** (/ˌproʊtərəˈzoʊɪk, <wbr />prɔː-, <wbr />-trə-/[73,74]) is a geological eon spanning the time from the appearance of oxygen in Earth's atmosphere to just before the proliferation of complex life (such as trilobites or corals) on the Earth. The name Proterozoic combines the two forms of ultimately Greek origin: *protero-* meaning "former, earlier", and *-zoic*, a suffix related to *zoe* "life". The Proterozoic Eon extended from 2500 Ma to 541 Ma (million years ago), and is the most recent part of the Precambrian "supereon".

The Proterozoic is the longest eon of the Earth's geologic time scale and it is subdivided into three geologic eras (from oldest to youngest): the Paleoproterozoic, Mesoproterozoic, and Neoproterozoic.

The well-identified events of this eon were the transition to an oxygenated atmosphere during the Paleoproterozoic; several glaciations, which produced the hypothesized Snowball Earth during the Cryogenian Period in the late Neoproterozoic Era; and the Ediacaran Period (635 to 541 Ma) which is characterized by the evolution of abundant soft-bodied multicellular organisms and provides us with the first obvious fossil evidence of life on earth.

The Proterozoic record

Life timeline

Axis scale: million years

🖐

Also see: *Human timeline* and *Nature timeline*

The geologic record of the Proterozoic Eon is more complete than that for the preceding Archean Eon. In contrast to the deep-water deposits of the Archean, the Proterozoic features many strata that were laid down in extensive shallow epicontinental seas; furthermore, many of those rocks are less metamorphosed than are Archean ones, and many are unaltered.[315] Studies of these rocks have shown that the eon continued the massive continental accretion that had begun late in the Archean Eon. The Proterozoic Eon also featured the first definitive supercontinent cycles and wholly modern mountain building activity (orogeny).[315–18, 329–32]

There is evidence that the first known glaciations occurred during the Proterozoic. The first began shortly after the beginning of the Proterozoic Eon, and evidence of at least four during the Neoproterozoic Era at the end of the Proterozoic Eon, possibly climaxing with the hypothesized Snowball Earth of the Sturtian and Marinoan glaciations.[320–1, 325]

The accumulation of oxygen

One of the most important events of the Proterozoic was the accumulation of oxygen in the Earth's atmosphere. Though oxygen is believed to have been released by photosynthesis as far back as Archean Eon, it could not build up to any significant degree until mineral sinks of unoxidized sulfur and iron had been filled. Until roughly 2.3 billion years ago, oxygen was probably only 1% to 2% of its current level.[323] The Banded iron formations, which provide most of the world's iron ore, are one mark of that mineral sink process. Their accumulation ceased after 1.9 billion years ago, after the iron in the oceans had all been oxidized.[324]

Red beds, which are colored by hematite, indicate an increase in atmospheric oxygen 2 billion years ago. Such massive iron oxide formations are not found in older rocks.[324] The oxygen buildup was probably due to two factors: a filling of the chemical sinks, and an increase in carbon burial, which sequestered organic compounds that would have otherwise been oxidized by the atmosphere.[325]

Subduction processes

The Proterozoic Eon was a very tectonically active period in the Earth's history. The late Archean Eon to Early Proterozoic Eon corresponds to a period of increasing crustal recycling, suggesting subduction. Evidence for this increased subduction activity comes from the abundance of old granites originating mostly after 2.6 Ga.[75] The appearance of eclogites, which metamorphic rocks created by high pressure (>1 GPa), are explained using a model that incorporates subduction. The lack of eclogites that date to the Archean Eon suggests that conditions at that time did not favor the formation of high grade metamorphism and therefore did not achieve the same levels of subduction as was occurring in the Proterozoic Eon. As a result of remelting of basaltic oceanic crust due to subduction, the cores of the first continents grew large enough to withstand the crustal recycling processes. The long-term tectonic stability of those cratons is why we find continental crust ranging up to a few billion years in age.[76] It is believed that 43% of modern continental crust was formed in the Proterozoic, 39% formed in the Archean, and only 18% in the Phanerozoic. Studies by Condie 2000Wikipedia:Citation needed and Rino et al. 2004Wikipedia:Citation needed suggest that crust production happened episodically. By isotopically calculating the ages of Proterozoic granitoids it was determined that there were several episodes of rapid increase in continental crust production. The reason for these pulses is unknown, but they seemed to have decreased in magnitude after every period.

Tectonic history (supercontinents)

Evidence of collision and rifting between continents raises the question as to what exactly were the movements of the Archean cratons composing Proterozoic continents. Paleomagnetic and geochronological dating mechanisms have allowed the deciphering of Precambrian Supereon tectonics. It is known that tectonic processes of the Proterozoic Eon resemble greatly the evidence of tectonic activity, such as orogenic belts or ophiolite complexes, we see today. Hence, most geologists would conclude that the Earth was active at that time. It is also commonly accepted that during the Precambrian, the Earth went through several supercontinent breakup and rebuilding cycles (Wilson cycle). In the late Proterozoic (most recent), the dominant supercontinent was Rodinia (~1000–750 Ma). It consisted of a series of continents attached to a central craton that forms the core of the North American Continent called Laurentia. An example of an orogeny (mountain building processes) associated with the construction of Rodinia is the Grenville orogeny located in Eastern North America. Rodinia formed after the breakup of the supercontinent Columbia and prior to the assemblage of the supercontinent Gondwana (~500 Ma). The defining orogenic event associated with the formation of Gondwana was the collision of Africa, South America, Antarctica and Australia forming the Pan-African orogeny.[77]

Columbia was dominant in the early-mid Proterozoic and not much is known about continental assemblages before then. There are a few plausible models that explain tectonics of the early Earth pre-Columbia, but the current most plausible theory is that prior to Columbia, there were only a few independent craton formations scattered around the Earth (not necessarily a supercontinent formation like Rodinia or Columbia).

Life

Stromatolites

South America

Western Namibia

The first advanced single-celled, eukaryotes and multi-cellular life, Francevillian Group Fossils, roughly coincides with the start of the accumulation of free oxygen. This may have been due to an increase in the oxidized nitrates that eukaryotes use, as opposed to cyanobacteria.[:325] It was also during the Proterozoic that the first symbiotic relationships between mitochondria (found in nearly all eukaryotes) and chloroplasts (found in plants and some protists only) and their hosts evolved.[:321-2]

The blossoming of eukaryotes such as acritarchs did not preclude the expansion of cyanobacteria; in fact, stromatolites reached their greatest abundance and diversity during the Proterozoic, peaking roughly 1200 million years ago.[:321-3]

Classically, the boundary between the Proterozoic and the Phanerozoic eons was set at the base of the Cambrian Period when the first fossils of animals including trilobites and archeocyathids appeared. In the second half of the

20th century, a number of fossil forms have been found in Proterozoic rocks, but the upper boundary of the Proterozoic has remained fixed at the base of the Cambrian, which is currently placed at 541 Ma.

External links

Wikimedia Commons has media related to *Proterozoic*.

- Palaeos.com: Proterozoic eon[78]

Phanerozoic Eon

Phanerozoic

Phanerozoic Eon
541–0 million years ago

The geological eons and eras

0 —
500
1000
1500
2000
2500
3000
3500
4000
4500

Scale: millions of years

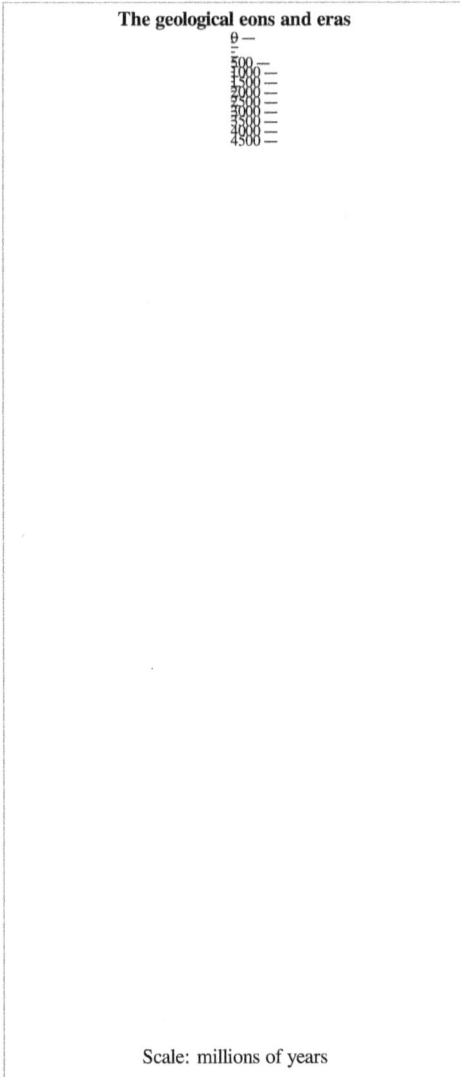

The **Phanerozoic** Eon[79,80]</ref> is the current geologic eon in the geologic time scale, and the one during which abundant animal and plant life has existed. It covers 541 million years to the present,[81] and began with the Cambrian Period when diverse hard-shelled animals first appeared. Its name was derived from the Ancient Greek words φανερός (*phanerós*) and ζωή (*zōḗ*), meaning *visible life*, since it was once believed that life began in the Cambrian, the first period of this eon. The term "Phanerozoic" was coined in 1930 by the American geologist George Halcott Chadwick (1876–1953). The time before the Phanerozoic, called the *Precambrian*, is now divided into the Hadean, Archaean and Proterozoic eons.

The time span of the Phanerozoic starts with what appears to be the rapid emergence of a number of animal phyla; the evolution of those phyla into diverse forms; the emergence and development of complex plants; the evolution of fish; the emergence of insects and tetrapods; and the development of modern fauna. Plant life on land appeared in the early Phanerozoic eon. During this time span, tectonic forces caused the continents to move and eventually collect into a single landmass known as Pangaea (the most recent supercontinent), which then separated into the current continental landmasses.

Proterozoic-Phanerozoic boundary

Life timeline

Axis scale: million years

Also see: *Human timeline* and *Nature timeline*

The Proterozoic-Phanerozoic boundary is at 541 million years ago. In the 19th century, the boundary was set at time of appearance of the first abundant animal (metazoan) fossils but several hundred groups (taxa) of metazoa of the earlier Proterozoic era have been identified since the systematic study of those forms started in the 1950s. Most geologists and paleontologists would probably set the Proterozoic-Phanerozoic boundary either at the classic point where the first trilobites and reef-building animals (archaeocyatha) such as corals and others appear; at the first appearance of a complex feeding burrow called *Treptichnus pedum*; or at the first appearance of a group of small, generally disarticulated, armored forms termed 'the small shelly fauna'. The three different dividing points are within a few million years of each other.

In the older literature, the term *Phanerozoic* is generally used as a label for the time period of interest to paleontologists, but that use of the term seems to be falling into disuse in more modern literature.

Eras of the Phanerozoic

The Phanerozoic is divided into three eras: the Paleozoic, Mesozoic, and Cenozoic, which are further subdivided into 12 periods. The Paleozoic features the rise of fish, amphibians and reptiles. The Mesozoic is ruled by the reptiles, and features the evolution of mammals, birds and more famously, dinosaurs. The Cenozoic is the time of the mammals, and more recently, humans.

Paleozoic Era

The Paleozoic is a time in Earth's history when complex life forms evolved, took their first breath of oxygen on dry land, and when the forerunners of all life on Earth began to diversify. There are six periods in the Paleozoic era: Cambrian, Ordovician, Silurian, Devonian, Carboniferous and Permian.

Cambrian Period

The Cambrian is the first period of the Paleozoic Era and starts from 541 to 485[82] million years ago. The Cambrian sparked a rapid expansion in evolution in an event known as the Cambrian Explosion during which the greatest number of creatures evolved in a single period in the history of Earth. Plants like algae evolved, and the fauna was dominated by armored arthropods, such as trilobites. Almost all marine phyla evolved in this period. During this time, the super-continent Pannotia began to break up, most of which later recombined into the super-continent Gondwana.

Figure 12: *Trilobites*

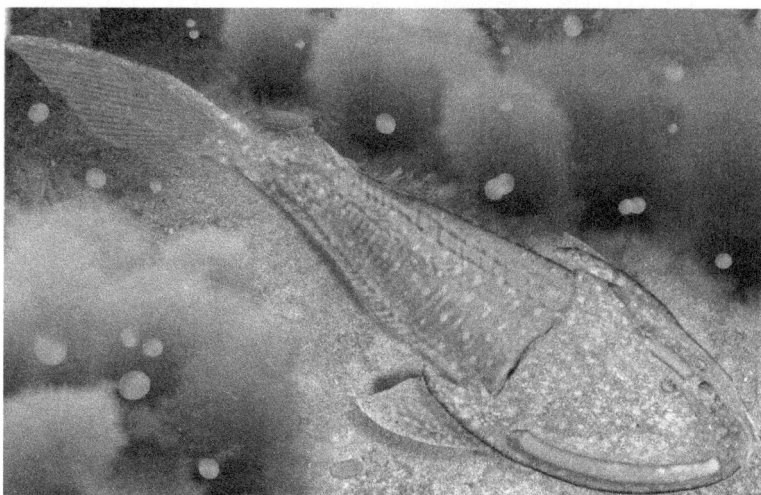

Figure 13: *Cephalaspis, a jawless fish*

Ordovician Period

The Ordovician spans from 485 million years to 440 million years ago. The Ordovician was a time in Earth's history in which many species still prevalent today evolved, such as primitive fish, cephalopods, and coral. The most common forms of life, however, were trilobites, snails and shellfish. More importantly, the first arthropods crept ashore to colonize Gondwana, a continent empty of animal life. By the end of the Ordovician, Gondwana had moved from the equator to the South Pole, and Laurentia had collided with Baltica, closing the Iapetus Ocean. The glaciation of Gondwana resulted in a major drop in sea level, killing off all life that had established along its coast. Glaciation caused a snowball Earth, leading to the Ordovician-Silurian extinction, during which 60% of marine invertebrates and 25% of families became extinct. This is considered the first mass extinction and the second deadliest in the history of Earth.

Silurian Period

The Silurian spans from 440 million years to 415 million years ago, which saw a warming from Snowball Earth. This period saw the mass evolution of fish, as jaw-less fish became more numerous, jawed fish evolved, and the first freshwater fish evolved, though arthropods, such as sea scorpions, remained the apex predators. Fully terrestrial life evolved, which included early arachnids, fungi, and centipedes. The evolution of vascular plants (Cooksonia) allowed plants to gain a foothold on land. These early terrestrial plants are the forerunners of all plant life on land. During this time, there were four continents: Gondwana (Africa, South America, Australia, Antarctica, India), Laurentia (North America with parts of Europe), Baltica (the rest of Europe), and Siberia (Northern Asia). The recent rise in sea levels provided new habitats for many new species.

Devonian Period

The Devonian spans from 415 million years to 360 million years ago. Also known as the "Age of the Fish", the Devonian features a huge diversification in fish, including armored fish like *Dunkleosteus* and lobe-finned fish which eventually evolved into the first tetrapods. On land, plant groups diversified incredibly in an event known as the Devonian Explosion during which the first trees evolved, as well as seeds. This event also allowed the diversification of arthropod life as they took advantage of the new habitat. The first amphibians also evolved, and the fish were now at the top of the food chain. Near the end of the Devonian, 70% of all species became extinct in an event known as the Late Devonian extinction, which is the second mass extinction known to have happened.

Figure 14: *Eogyrinus (an amphibian) of the Carboniferous*

Figure 15: *Dimetrodon*

Carboniferous Period

The Carboniferous spans from 360 million to 300 million years ago. During this period, average global temperatures were exceedingly high: the early Carboniferous averaged at about 20 degrees Celsius (but cooled to 10 degrees during the Middle Carboniferous). Tropical swamps dominated the Earth, and the large amounts of trees created much of the carbon that became coal deposits (hence the name Carboniferous). The high oxygen levels caused by these swamps allowed massive arthropods, normally limited in size by their respiratory systems, to proliferate. Perhaps the most important evolutionary development of the time was the evolution of amniotic eggs, which allowed amphibians to move farther inland and remain the dominant vertebrates throughout the period. Also, the first reptiles and synapsids evolved in the swamps. Throughout the Carboniferous, there was a cooling pattern, which eventually led to the glaciation of Gondwana as much of it was situated around the south pole, in an event known as the Permo-Carboniferous glaciation or the Carboniferous Rainforest Collapse.

Permian Period

The Permian spans from 300 million to 250 million years ago and was the last period of the Paleozoic Era. At its beginning, all continents came together to form the super-continent Pangaea, surrounded by one ocean called Panthalassa. The Earth was very dry during this time, with harsh seasons, as the climate of the interior of Pangaea wasn't regulated by large bodies of water. Reptiles and synapsids flourished in the new dry climate. Creatures such as *Dimetrodon* and *Edaphosaurus* ruled the new continent. The first conifers evolved, then dominated the terrestrial landscape. Nearing the end of the period, *Scutosaurus* and gorgonopsids filled the empty desert. Eventually, they disappeared, along with 95% of all life on Earth in an event simply known as "the Great Dying", the world's third mass extinction event and the largest in its history.

Mesozoic Era

The Mesozoic ranges from 252 million to 66 million years ago. Also known as "the Age of the dinosaurs", the Mesozoic features the rise of reptiles on their 150 million year conquest of the Earth on the land, in the seas, and in the air. There are three periods in the Mesozoic: Triassic, Jurassic, and Cretaceous.

Triassic Period

The Triassic ranges from 250 million to 200 million years ago. The Triassic is a desolate transitional time in Earth's history between the Permian Extinction and the lush Jurassic Period. It has three major epochs: Early Triassic, Middle Triassic and Late Triassic.

The Early Triassic lasted between 250 million to 247 million years ago, and was dominated by deserts as Pangaea had not yet broken up, thus the interior was arid. The Earth had just witnessed a massive die-off in which 95% of all life became extinct. The most common life on Earth were *Lystrosaurus*,

Figure 16: *Plateosaurus (a prosauropod)*

labyrinthodonts, and *Euparkeria* along with many other creatures that managed to survive the Great Dying. Temnospondyli evolved during this time and would be the dominant predator for much of the Triassic.

The Middle Triassic spans from 247 million to 237 million years ago. The Middle Triassic featured the beginnings of the breakup of Pangaea, and the beginning of the Tethys Sea. The ecosystem had recovered from the devastation of the Great Dying. Phytoplankton, coral, and crustaceans all had recovered, and the reptiles began increasing in size. New aquatic reptiles, such as ichthyosaurs and nothosaurs, evolved. Meanwhile, on land, pine forests flourished, as well as mosquitoes and fruit flies. The first ancient crocodilians evolved, which sparked competition with the large amphibians that had long ruled the freshwater world.

The Late Triassic spans from 237 million to 200 million years ago. Following the bloom of the Middle Triassic, the Late Triassic featured frequent rises of temperature, as well as moderate precipitation (10-20 inches per year). The recent warming led to a boom of reptilian evolution on land as the first true dinosaurs evolved, as well as pterosaurs. The climactic change, however, resulted in a large die-out known as the Triassic-Jurassic extinction event, in which all archosaurs (excluding ancient crocodiles), synapsids, and almost all large amphibians became extinct, as well as 34% of marine life in the fourth mass extinction event. The extinction's cause is debated.

Figure 17: *Rhamphorhynchus*

Jurassic Period

The Jurassic ranges from 200 million to 145 million years ago, and features three major epochs: Early Jurassic, Middle Jurassic, and Late Jurassic.

The Early Jurassic Epoch spans from 200 million to 175 million years ago. The climate was much more humid than the Triassic, and as a result, the world was very tropical. In the oceans, plesiosaurs, ichthyosaurs and ammonites dominated the seas. On land, dinosaurs and other reptiles dominated the land, with species such as *Dilophosaurus* at the apex. The first true crocodiles evolved, pushing the large amphibians to near extinction. The reptiles rose to rule the world. Meanwhile, the first true mammals evolved, but never exceeded the height of a shrew.

The Middle Jurassic Epoch spans from 175 million to 163 million years ago. During this epoch, reptiles flourished as huge herds of sauropods, such as *Brachiosaurus* and *Diplodicus*, filled the fern prairies of the Middle Jurassic. Many other predators rose as well, such as *Allosaurus*. Conifer forests made up a large portion of the world's forests. In the oceans, plesiosaurs were quite common, and ichthyosaurs were flourishing. This epoch was the peak of the reptiles.

The Late Jurassic Epoch spans from 163 million to 145 million years ago. The Late Jurassic featured a massive extinction of sauropods and ichthyosaurs due to the separation of Pangaea into Laurasia and Gondwana in an extinction known as the Jurassic-Cretaceous extinction. Sea levels rose, destroying fern prairies and creating shallows. Ichthyosaurs became extinct whereas

Figure 18: *Artist's 1901 depiction of a Stegosaurus (inaccurately portrayed with a dragging tail).*

sauropods, as a whole, did not; in fact, some species, like *Titanosaurus*, lived until the K-T extinction. The increase in sea-levels opened up the Atlantic sea way which would continue to get larger over time. The divided world would give opportunity for the diversification of new dinosaurs.

Cretaceous Period

The Cretaceous is the longest period in the Mesozoic, spans from 145 million to 66 million years ago, and is divided into two epochs: Early Cretaceous, and Late Cretaceous.

The Early Cretaceous Epoch spans from 145 million to 100 million years ago. The Early Cretaceous saw the expansion of seaways, and as a result, the decline and extinction of sauropods (except in South America). Many coastal shallows were created, and that caused ichthyosaurs to die out. Mosasaurs evolved to replace them as apex species of the seas. Some island-hopping dinosaurs, like *Eustreptospondylus*, evolved to cope with the coastal shallows and small islands of ancient Europe. Other dinosaurs, such as *Carcharodontosaurus* and *Spinosaurus*, rose to fill the empty space that the Jurassic-Cretaceous extinction had created. Of the most successful would be the *Iguanodon* which spread to every continent. Seasons came back into effect and the poles grew seasonally colder. Dinosaurs such as the *Leaellynasaura* inhabited the polar forests year-round, while many dinosaurs, such as the *Muttaburrasaurus*, migrated there during summer . Since it was too cold for crocodiles, it was

Figure 19: *Tylosaurus (a mosasaur) hunting Xiphactinus*

the last stronghold for large amphibians, such as the *Koolasuchus*. Pterosaurs grew larger as species like *Tapejara* and *Ornithocheirus* evolved. More importantly, the first true birds evolved sparking competition between them and the pterosaurs.

The Late Cretaceous Epoch spans from 100 million to 65 million years ago. The Late Cretaceous featured a cooling trend that would continue into the Cenozoic Era. Eventually, tropical ecology was restricted to the equator and areas beyond the tropic lines featured extreme seasonal changes of weather. Dinosaurs still thrived as new species such as *Tyrannosaurus*, *Ankylosaurus*, *Triceratops* and Hadrosaurs dominated the food web. Pterosaurs, however, were going into a decline as birds took to the skies. The last pterosaur to die off was *Quetzalcoatlus*. Marsupials evolved within the large conifer forests as scavengers. In the oceans, Mosasaurs ruled the seas to fill the role of the ichthyosaurs, and huge plesiosaurs, such as *Elasmosaurus*, evolved. Also, the first flowering plants evolved. At the end of the Cretaceous, the Deccan Traps and other volcanic eruptions were poisoning the atmosphere. As this was continued, it is thought that a large meteor smashed into Earth, creating the Chicxulub Crater creating the event known as the K-T Extinction, the fifth and most recent mass extinction event, during which 75% of life on Earth became extinct, including all non-avian dinosaurs. Every living thing with a body mass over 10 kilograms became extinct, and the age of the dinosaurs came to an end.

Figure 20: *Basilosaurus (a whale, despite the name)*

Cenozoic Era

The Cenozoic featured the rise of mammals as the dominant class of animals, as the end of the age of the dinosaurs left significant evolutionary vacuums. There are three divisions of the Cenozoic: Paleogene, Neogene and Quaternary.

Paleogene Period

The Paleogene spans from the extinction of the dinosaurs, some 66 million years ago, to the dawn of the Neogene 23 million years ago. It features three epochs: Paleocene, Eocene and Oligocene.

The Paleocene Epoch began with the K-T extinction event caused by the impact of a metorite in the area of present-day Yucatan Peninsula and caused the destruction of 75% of all species on Earth. The Early Paleocene saw the recovery of the Earth from that event. The continents began to take their modern shape, but all continents (and India) were separated from each other. Afro-Eurasia was separated by the Tethys Sea, and the Americas were separated by the strait of Panama, as the Isthmus of Panama had not yet formed. This epoch featured a general warming trend, and jungles eventually reached the poles. The oceans were dominated by sharks as the large reptiles that had once ruled became extinct. Archaic mammals, such as creodonts and early primates that evolved during the Mesozoic filled the world. During this time there were no land creatures over 10 kilograms. Mammals were still quite small.

The Eocene Epoch ranged from 56 million to 34 million years ago. In the early Eocene, land animals were small and living in cramped jungles, much like the Paleocene. None had a mass over 10 kilograms. Among them were early primates, whales and horses along with many other early forms of mammals. At the top of the food chains were huge birds, such as *Gastornis*. It is the only time in recorded history that birds ruled the world (excluding their ancestors, the dinosaurs). The temperature was 30 degrees Celsius with little temperature gradient from pole to pole. In the Middle Eocene Epoch, the circum-Antarctic current between Australia and Antarctica formed which disrupted ocean currents worldwide, resulting in global cooling, and caused the jungles to shrink. This allowed mammals to grow; some such as whales to mammoth proportions, which were, by now, almost fully aquatic. Mammals like *Andrewsarchus* were now at the top of the food-chain and sharks were replaced by *Basilosaurus*, whales, as rulers of the seas. The late Eocene Epoch saw the rebirth of seasons, which caused the expansion of savanna-like areas, along with the evolution of grass.

The Oligocene Epoch spans from 33 million to 23 million years ago. The Oligocene featured the expansion of grass which had led to many new species to take advantage, including the first elephants, cats, dogs, marsupials and many other species still prevalent today. Many other species of plants evolved during this epoch also, such as the evergreen trees. The long term cooling continued and seasonal rains patterns established. Mammals continued to grow larger. *Paraceratherium*, the largest land mammal to ever live evolved during this epoch, along with many other perissodactyls in an event known as the Eocene–Oligocene extinction event (Grand Coupure).

Neogene Period

The Neogene spans from 23.03 million to 2.58 million years ago. It features 2 epochs: the Miocene, and the Pliocene.

The Miocene spans from 23.03 to 5.333 million years ago and is a period in which grass spread further across, effectively dominating a large portion of the world, diminishing forests in the process. Kelp forests evolved, leading to the evolution of new species, such as sea otters. During this time, perissodactyla thrived, and evolved into many different varieties. Alongside them were the apes, which evolved into a 30 species. Overall, arid and mountainous land dominated most of the world, as did grazers. The Tethys Sea finally closed with the creation of the Arabian Peninsula and in its wake left the Black, Red, Mediterranean and Caspian Seas. This only increased aridity. Many new plants evolved, and 95% of modern seed plants evolved in the mid-Miocene.

Figure 21: *Animals of the Miocene (Chalicotherium, Hyenadon, entelodont)*

The Pliocene lasted from 5.333 to 2.58 million years ago. The Pliocene featured dramatic climactic changes, which ultimately led to modern species and plants. The Mediterranean Sea dried up for several million years. Along with these major geological events, *Australopithecus* evolved in Africa, beginning the human branch. The isthmus of Panama formed, and animals migrated between North and South America, wreaking havoc on the local ecology. Climatic changes brought savannas that are still continuing to spread across the world, Indian monsoons, deserts in East Asia, and the beginnings of the Sahara desert. The Earth's continents and seas moved into their present shapes. The world map has not changed much since, save for changes brought about by the glaciations of the Quaternary, such as the Great Lakes.

Quaternary Period

The Quaternary spans from 2.58 million years ago to present day, and is the shortest geological period in the Phanerozoic Eon. It features modern animals, and dramatic changes in the climate. It is divided into two epochs: the Pleistocene and the Holocene.

The Pleistocene lasted from 2.58 million to 11,700 years ago. This epoch was marked by ice ages as a result of the cooling trend that started in the Mid-Eocene. There were at least four separate glaciation periods marked by the advance of ice caps as far south as 40 degrees N latitude in mountainous

Figure 22: *Megafauna of the Pleistocene (mammoths, cave lions, woolly rhino, reindeer, horses)*

areas. Meanwhile, Africa experienced a trend of desiccation which resulted in the creation of the Sahara, Namib, and Kalahari deserts. Many animals evolved including mammoths, giant ground sloths, dire wolves, saber-toothed cats, and most famously *Homo sapiens*. 100,000 years ago marked the end of one of the worst droughts of Africa, and led to the expansion of primitive man. As the Pleistocene drew to a close, a major extinction wiped out much of the world's megafauna, including some of the hominid species, such as Neanderthals. All the continents were affected, but Africa to a lesser extent. That continent retains many large animals, such as hippos.

The Holocene began 11,700 years ago and lasts until to present day. All recorded history and "the history of the world" lies within the boundaries of the Holocene epoch. Human activity is blamed for a mass extinction that began roughly 10,000 years ago, though the species becoming extinct have only been recorded since the Industrial Revolution. This is sometimes referred to as the "Sixth Extinction". More than 322 species have become extinct due to human activity since the Industrial Revolution.

Biodiversity

It has been demonstrated that changes in biodiversity through the Phanerozoic correlate much better with the hyperbolic model (widely used in demography and macrosociology) than with exponential and logistic models (traditionally used in population biology and extensively applied to fossil biodiversity as well). The latter models imply that changes in diversity are guided by a first-order positive feedback (more ancestors, more descendants) or a negative feedback that arises from resource limitation, or both. The hyperbolic

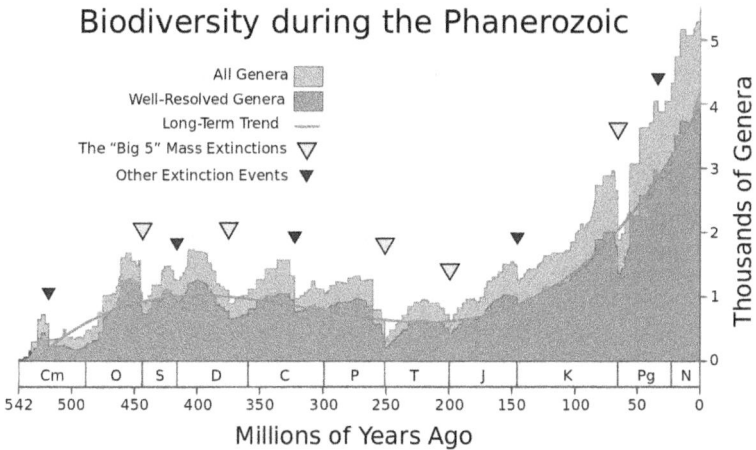

Figure 23: *During the Phanerozoic, biodiversity shows a steady but not monotonic increase from near zero to several thousands of genera.*

model implies a second-order positive feedback. The hyperbolic pattern of the human population growth arises from a second-order positive feedback, caused by the interaction of the population size and the rate of technological growth.[83] The character of biodiversity growth in the Phanerozoic Eon can be similarly accounted for by a feedback between the diversity and community structure complexity. It is suggested that the similarity between the curves of biodiversity and human population probably comes from the fact that both are derived from the superposition on the hyperbolic trend of cyclical and random dynamics.

References

- Markov, Alexander V.; Korotayev, Andrey V. (2007). "Phanerozoic marine biodiversity follows a hyperbolic trend"[84]. *Palaeoworld*. **16** (4): 311–318. doi: 10.1016/j.palwor.2007.01.002[85].
- Miller, K. G.; Kominz, M. A.; Browning, J. V.; Wright, J. D.; Mountain, G. S.; Katz, M. E.; Sugarman, P. J.; Cramer, B. S.; Christie-Blick, N; Pekar, S. F.; et al. (2005). "The Phanerozoic record of global sea-level change"[86] (PDF). *Science*. **310** (5752): 1293–1298. Bibcode: 2005Sci...310.1293M[87]. doi: 10.1126/science.1116412[88]. PMID 16311326[89].

External links

- 🌐 Media related to Phanerozoic at Wikimedia Commons

Paleozoic

Paleozoic Era
541 - 251.902 million years ago

Key events in the Paleozoic

Phanerozoic Eon
Proterozoic Eon
An approximate timescale of key Paleozoic events.
Axis scale: millions of years ago.

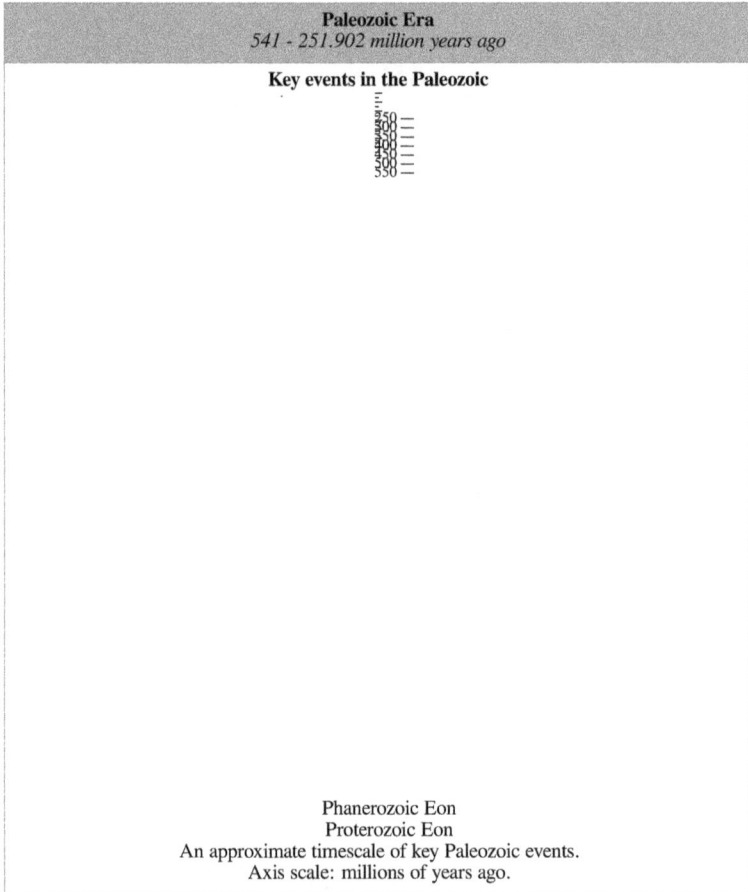

The **Paleozoic** (or **Palaeozoic**) **Era** (/ˌpeɪliəˈzoʊɪk, <wbr /> ˌpæ-/;[90,91] from the Greek *palaios* (παλαιός), "old" and *zoe* (ζωή), "life", meaning "ancient life"[92]) is the earliest of three geologic eras of the Phanerozoic Eon. It is the longest of the Phanerozoic eras, lasting from 541 to 251.902[93] million years ago, and is subdivided into six geologic periods (from oldest to youngest): the Cambrian, Ordovician, Silurian, Devonian, Carboniferous, and Permian. The

Paleozoic comes after the Neoproterozoic Era of the Proterozoic Eon and is followed by the Mesozoic Era.

The Paleozoic was a time of dramatic geological, climatic, and evolutionary change. The Cambrian witnessed the most rapid and widespread diversification of life in Earth's history, known as the Cambrian explosion, in which most modern phyla first appeared. Fish, arthropods, amphibians, anapsids, synapsids, euryapsids, and diapsids all evolved during the Paleozoic. Life began in the ocean but eventually transitioned onto land, and by the late Paleozoic, it was dominated by various forms of organisms. Great forests of primitive plants covered the continents, many of which formed the coal beds of Europe and eastern North America. Towards the end of the era, large, sophisticated diapsids and synapsids were dominant and the first modern plants (conifers) appeared.

The Paleozoic Era ended with the largest extinction event in the history of Earth, the Permian–Triassic extinction event. The effects of this catastrophe were so devastating that it took life on land 30 million years into the Mesozoic Era to recover. Recovery of life in the sea may have been much faster.[94]

Geology

The Paleozoic era began and ended with supercontinents and in between were the rise of mountains along the continental margins, and flooding and draining of shallow seas between.Wikipedia:Please clarify At its start, the supercontinent Pannotia broke up. Paleoclimatic studies and evidence of glaciers indicate that central Africa was most likely in the polar regions during the early Paleozoic. During the early Paleozoic, the huge continent Gondwana (510[95] million years ago) formed or was forming. By mid-Paleozoic, the collision of North America and Europe produced the Acadian-Caledonian uplifts, and a subduction plate uplifted eastern Australia. By the late Paleozoic, continental collisions formed the supercontinent of Pangaea and resulted in some of the great mountain chains, including the Appalachians, Ural Mountains, and mountains of Tasmania.

Periods of the Paleozoic Era

There are six periods in the Paleozoic Era: Cambrian, Ordovician, Silurian, Devonian, Carboniferous (alternatively subdivided into the Mississippian Period and the Pennsylvanian Period), and the Permian.[96]

Figure 24: *Trilobites*

Cambrian Period

The Cambrian spans from 541 million years to 485 million years and is the
first period of the Paleozoic era of the Phanerozoic. The Cambrian marked
a boom in evolution in an event known as the Cambrian explosion in which
the largest number of creatures evolved in any single period of the history of
the Earth. Creatures like algae evolved, but the most ubiquitous of that period
were the armored arthropods, like trilobites. Almost all marine phyla evolved
in this period. During this time, the supercontinent Pannotia begins to break
up, most of which later became the supercontinent Gondwana.

Ordovician Period

The Ordovician spanned from approximately 485 million years to approximately 443 million years ago. The Ordovician was a time in Earth's history
in which many of the biological classes still prevalent today evolved, such as
primitive fish, cephalopods, and coral. The most common forms of life, however, were trilobites, snails and shellfish. More importantly, the first arthropods
went ashore to colonize the empty continent of Gondwana. By the end of the
Ordovician, Gondwana was at the south pole, early North America had collided with Europe, closing the Atlantic Ocean. Glaciation of Africa resulted in
a major drop in sea level, killing off all life that had established along coastal
Gondwana. Glaciation may have caused the Ordovician–Silurian extinction
events, in which 60% of marine invertebrates and 25% of families became extinct, and is considered the first mass extinction event and the second deadliest.

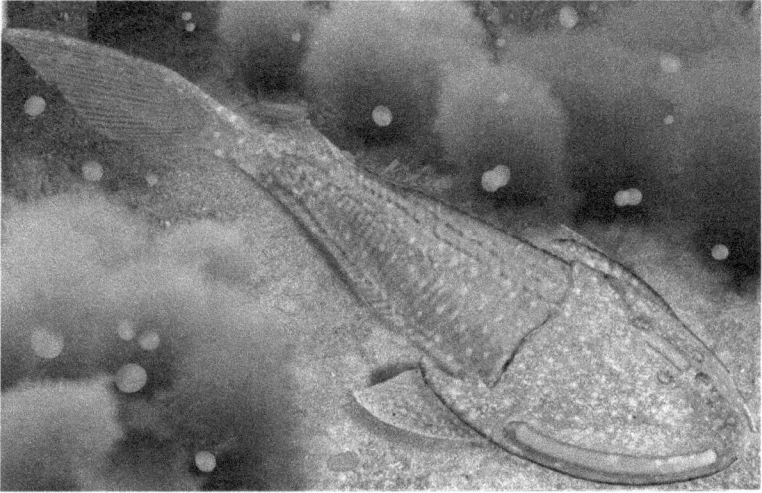

Figure 25: *Cephalaspis (a jawless fish)*

Silurian Period

The Silurian spanned from 443 to 416 million years ago. The Silurian saw the rejuvenation of life as the Earth recovered from the previous glaciation. This period saw the mass evolution of fish, as jawless fish became more numerous, jawed fish evolved, and the first freshwater fish evolved, though arthropods, such as sea scorpions, were still apex predators. Fully terrestrial life evolved, including early arachnids, fungi, and centipedes. The evolution of vascular plants (Cooksonia) allowed plants to gain a foothold on land. These early plants are the forerunners of all plant life on land. During this time, there were four continents: Gondwana (Africa, South America, Australia, Antarctica, Siberia), Laurentia (North America), Baltica (Northern Europe), and Avalonia (Western Europe). The recent rise in sea levels allowed many new species to thrive in water.

Devonian Period

The Devonian spanned from 416 million years to 359 million years. Also known as "The Age of the Fish", the Devonian featured a huge diversification of fish, including armored fish like *Dunkleosteus* and lobe-finned fish which eventually evolved into the first tetrapods. On land, plant groups diversified incredibly in an event known as the Devonian Explosion when plants made lignin allowing taller growth and vascular tissue: the first trees evolved, as well as seeds. This event also diversified arthropod life, by providing them new habitats. The first amphibians also evolved, and the fish were now at the top

Figure 26: *Eogyrinus (an amphibian) of the Carboniferous*

of the food chain. Near the end of the Devonian, 70% of all species became extinct in an event known as the Late Devonian extinction and was the second mass extinction event the world has seen.

Carboniferous Period

The Carboniferous spanned from 359 million to 299 million years. During this time, average global temperatures were exceedingly high; the early Carboniferous averaged at about 20 degrees Celsius (but cooled to 10 °C during the Middle Carboniferous). Tropical swamps dominated the Earth, and the lignin stiffened trees grew to greater heights and number. As the bacteria and fungi capable of eating the lignin had not yet evolved, their remains were left buried, which created much of the carbon that became the coal deposits of today (hence the name "Carboniferous"). Perhaps the most important evolutionary development of the time was the evolution of amniotic eggs, which allowed amphibians to move farther inland and remain the dominant vertebrates for the duration of this period. Also, the first reptiles and synapsids evolved in the swamps. Throughout the Carboniferous, there was a cooling trend, which led to the Permo-Carboniferous glaciation or the Carboniferous Rainforest Collapse. Gondwana was glaciated as much of it was situated around the south pole.

Figure 27: *Synapsid: Dimetrodon*

Permian Period

The Permian spanned from 299 to 252 million years ago and was the last period of the Paleozoic Era. At the beginning of this period, all continents joined together to form the supercontinent Pangaea, which was encircled by one ocean called Panthalassa. The land mass was very dry during this time, with harsh seasons, as the climate of the interior of Pangaea was not regulated by large bodies of water. Diapsids and synapsids flourished in the new dry climate. Creatures such as Dimetrodon and Edaphosaurus ruled the new continent. The first conifers evolved, and dominated the terrestrial landscape. Near the end of the Permian, however, Pangaea grew drier. The interior was desert, and new species such as Scutosaurus and Gorgonopsids filled it. Eventually they disappeared, along with 95% of all life on Earth, in a cataclysm known as "The Great Dying", the third and most severe mass extinction.

Tectonic activity

Geologically, the Paleozoic started shortly after the breakup of the supercontinent Pannotia. Throughout the early Paleozoic, that landmass was broken into a substantial number of continents. Towards the end of the era, the continents gathered together into a supercontinent called Pangaea, which included most of the Earth's land area.

Climate

The earliest two periods, the Ordovician and Silurian, were warm greenhouse periods, with the highest sea levels of the Paleozoic (200 m above today's); the warm climate was interrupted only by a 30^{97} million year cool period, the Early Palaeozoic Icehouse, culminating in the Huronian glaciation, 445^{98} million years ago at the end of the Ordovician.

The early Cambrian climate was probably moderate at first, becoming warmer over the course of the Cambrian, as the second-greatest sustained sea level rise in the Phanerozoic got underway. However, as if to offset this trend, Gondwana moved south, so that, in Ordovician time, most of West Gondwana (Africa and South America) lay directly over the South Pole. The early Paleozoic climate was also strongly zonal, with the result that the "climate", in an abstract sense, became warmer, but the living space of most organisms of the time—the continental shelf marine environment—became steadily colder. However, Baltica (Northern Europe and Russia) and Laurentia (eastern North America and Greenland) remained in the tropical zone, while China and Australia lay in waters which were at least temperate. The early Paleozoic ended, rather abruptly, with the short, but apparently severe, late Ordovician ice age. This cold spell caused the second-greatest mass extinction of Phanerozoic time. Over time, the warmer weather moved into the Paleozoic Era.

The middle Paleozoic was a time of considerable stability. Sea levels had dropped coincident with the ice age, but slowly recovered over the course of the Silurian and Devonian. The slow merger of Baltica and Laurentia, and the northward movement of bits and pieces of Gondwana created numerous new regions of relatively warm, shallow sea floor. As plants took hold on the continental margins, oxygen levels increased and carbon dioxide dropped, although much less dramatically. The north–south temperature gradient also seems to have moderated, or metazoan life simply became hardier, or both. At any event, the far southern continental margins of Antarctica and West Gondwana became increasingly less barren. The Devonian ended with a series of turnover pulses which killed off much of middle Paleozoic vertebrate life, without noticeably reducing species diversity overall.

There are many unanswered questions about the late Paleozoic. The Mississippian (early Carboniferous Period) began with a spike in atmospheric oxygen, while carbon dioxide plummeted to new lows. This destabilized the climate and led to one, and perhaps two, ice ages during the Carboniferous. These were far more severe than the brief Late Ordovician ice age; but, this time, the effects on world biota were inconsequential. By the Cisuralian Epoch, both oxygen and carbon dioxide had recovered to more normal levels. On the other hand, the assembly of Pangaea created huge arid inland areas subject to

Figure 28: *An artist's impression of early land plants*

temperature extremes. The Lopingian Epoch is associated with falling sea levels, increased carbon dioxide and general climatic deterioration, culminating in the devastation of the Permian extinction.

Flora

While macroscopic plant life appeared early in the Paleozoic Era and possibly late in the Neoproterozoic Era of the earlier eon. Plants mostly remained aquatic until sometime in the Silurian and Devonian Periods, about 420 million years ago, when they began to transition onto dry land. Terrestrial flora reached its climax in the Carboniferous, when towering lycopsid rainforests dominated the tropical belt of Euramerica. Climate change caused the Carboniferous Rainforest Collapse which fragmented this habitat, diminishing the diversity of plant life in the late Carboniferous and Permian.

Fauna

A noteworthy feature of Paleozoic life is the sudden appearance of nearly all of the invertebrate animal phyla in great abundance at the beginning of the Cambrian. The first vertebrates appeared in the form of primitive fish, which greatly diversified in the Silurian and Devonian Periods. The first animals to venture onto dry land were the arthropods. Some fish had lungs, and powerful bony fins that in the late Devonian, 367.5 million years ago, allowed them to crawl onto land. The bones in their fins eventually evolved into legs and they became the first tetrapods, 390[99] million years ago, and began to develop lungs. Amphibians were the dominant tetrapods until the mid-Carboniferous, when climate change greatly reduced their diversity. Later, reptiles prospered and continued to increase in number and variety by the late Permian.

Further reading

<templatestyles src="Template:Refbegin/styles.css" />

- *British Palaeozoic Fossils*, 1975, The Natural History Museum, London.
- "International Commission on Stratigraphy (ICS)"[100]. *Home Page*. Retrieved September 19, 2005.

External links

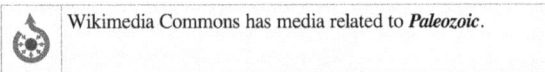
Wikimedia Commons has media related to *Paleozoic*.

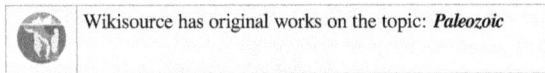
Wikisource has original works on the topic: *Paleozoic*

- 60+ images of Paleozoic Foraminifera[101]

Cambrian

Cambrian Period	
541–485.4 million years ago Pre€€ OSD C P T J K PgN	
Mean atmospheric O ₂ content over period duration	c. 12.5 vol %[102,103] (63 % of modern level
Mean atmospheric CO ₂ content over period duration	c. 4500 ppm[104] (16 times pre-industrial level)
Mean surface temperature over period duration	c. 21 °C[105] (7 °C above modern level)
Sea level (above present day)	Rising steadily from 30m to 90m

Key events in the Cambrian

‾
‾
‾490 —
‾500 —
‾510 —
‾520 —
‾530 —
‾540 —
550 —

The **Cambrian** Period (/'kæmbriən/ or /'keɪmbriən/) was the first geological period of the Paleozoic Era, and of the Phanerozoic Eon. The Cambrian lasted 55.6 million years from the end of the preceding Ediacaran Period 541 million years ago (mya) to the beginning of the Ordovician Period 485.4 mya. Its subdivisions, and its base, are somewhat in flux. The period was established (as "Cambrian series") by Adam Sedgwick, who named it after Cambria, the Latinised form of *Cymru*, the Welsh name for Wales, where Britain's Cambrian rocks are best exposed.[106] The Cambrian is unique in its unusually high proportion of lagerstätte sedimentary deposits, sites of exceptional preservation where "soft" parts of organisms are preserved as well as their more resistant shells. As a result, our understanding of the Cambrian biology surpasses that of some later periods.

The Cambrian marked a profound change in life on Earth; prior to the Cambrian, the majority of living organisms on the whole were small, unicellular and simple; the Precambrian *Charnia* being exceptional. Complex, multicellular organisms gradually became more common in the millions of years immediately preceding the Cambrian, but it was not until this period that mineralized—hence readily fossilized—organisms became common. The rapid diversification of lifeforms in the Cambrian, known as the Cambrian explosion, produced the first representatives of all modern animal phyla. Phylogenetic analysis has supported the view that during the Cambrian radiation, metazoa (animals) evolved monophyletically from a single common ancestor: flagellated colonial protists similar to modern choanoflagellates.

Although diverse life forms prospered in the oceans, the land is thought to have been comparatively barren—with nothing more complex than a microbial soil crust[107] and a few molluscs that emerged to browse on the microbial biofilm known to have been present. Most of the continents were probably dry and rocky due to a lack of vegetation. Shallow seas flanked the margins of several continents created during the breakup of the supercontinent Pannotia. The seas were relatively warm, and polar ice was absent for much of the period.

Stratigraphy

Despite the long recognition of its distinction from younger Ordovician rocks and older Precambrian rocks, it was not until 1994 that the Cambrian system/period was internationally ratified. The base of the Cambrian lies atop a complex assemblage of trace fossils known as the *Treptichnus pedum* assemblage.[108] The use of *Treptichnus pedum*, a reference ichnofossil to mark the lower boundary of the Cambrian, is difficult as the occurrence of very similar trace fossils belonging to the Treptichnids group are found well below the

T. pedum in Namibia, Spain and Newfoundland, and possibly, in the western USA. The stratigraphic range of *T. pedum* overlaps the range of the Ediacaran fossils in Namibia, and probably in Spain.[109,110]

Subdivisions

The Cambrian Period followed the Ediacaran Period and was followed by the Ordovician Period. The Cambrian is divided into four epochs (series) and ten ages (stages). Currently only three series and six stages are named and have a GSSP (an internationally agreed-upon stratigraphic reference point).

Because the international stratigraphic subdivision is not yet complete, many local subdivisions are still widely used. In some of these subdivisions the Cambrian is divided into three series (epochs) with locally differing names – the Early Cambrian (Caerfai or Waucoban, $541^{111} \pm 1.0$ to $509^{112} \pm 1.7$ mya), Middle Cambrian (St Davids or Albertan, $509^{112} \pm 1.0$ to $497^{113} \pm 1.7$ mya) and Furongian ($497^{113} \pm 1.0$ to $485.4^{114} \pm 1.7$ mya; also known as Late Cambrian, Merioneth or Croixan). Rocks of these epochs are referred to as belonging to the Lower, Middle, or Upper Cambrian.

Trilobite zones allow biostratigraphic correlation in the Cambrian.

Each of the local series is divided into several stages. The Cambrian is divided into several regional faunal stages of which the Russian-Kazakhian system is most used in international parlance:

		Chinese	North American	Russian-Kazakhian	Australian	Regional
C a m b r i a n	Furongian		Ibexian (part)	Ayusokkanian	Datsonian	Dolgellian (Trempealeauan, Fengshanian)
				Payntonian		
			Sunwaptan	Sakian	Iverian	Ffestiniogian (Franconian, Changshanian)
			Steptoan	Aksayan	Idamean	Maentwrogian (Dresbachian)
			Marjuman	Batyrbayan	Mindyallan	
	Miaolingian	Maozhangian		Mayan	Boomerangian	
		Zuzhuangian	Delamaran	Amgan	Undillian	
		Zhungxian			Florian	
					Templetonian	
			Dyeran		Ordian	

Cambrian Series 2	Longwang-mioan		Toyonian		Lenian
	Changlang-puan	Mon-tezuman	Botomian		
	Qungzusian		Atdabanian		
Terreneu-vian					
	Meishuchuan Jinningian	Placen-tian	Tommotian Nemakit-Daldynian*		Cordubian
Precambrian	Sinian	Hadry-nian	Nemakit-Daldynian* Sakharan	Adeladean	

*In Russian scientific thought the lower boundary of the Cambrian is suggested to be defined at the base of the Tommotian Stage which is characterized by diversification and global distribution of organisms with mineral skeletons and the appearance of the first Archaeocyath bioherms.

Dating the Cambrian

The International Commission on Stratigraphy list the Cambrian period as beginning at 541[111] million years ago and ending at 485.4[114] million years ago.

The lower boundary of the Cambrian was originally held to represent the first appearance of complex life, represented by trilobites. The recognition of small shelly fossils before the first trilobites, and Ediacara biota substantially earlier, led to calls for a more precisely defined base to the Cambrian period.

After decades of careful consideration, a continuous sedimentary sequence at Fortune Head, Newfoundland was settled upon as a formal base of the Cambrian period, which was to be correlated worldwide by the earliest appearance of *Treptichnus pedum*. Discovery of this fossil a few metres below the GSSP led to the refinement of this statement, and it is the *T. pedum* ichnofossil assemblage that is now formally used to correlate the base of the Cambrian.

This formal designation allowed radiometric dates to be obtained from samples across the globe that corresponded to the base of the Cambrian. Early dates of 570[115] million years ago quickly gained favour, though the methods used to obtain this number are now considered to be unsuitable and inaccurate. A more precise date using modern radiometric dating yield a date of 541[111] ± 0.3 million years ago. The ash horizon in Oman from which this date was recovered corresponds to a marked fall in the abundance of carbon-13 that correlates to equivalent excursions elsewhere in the world, and to the disappearance of distinctive Ediacaran fossils (*Namacalathus*, *Cloudina*). Nevertheless, there are arguments that the dated horizon in Oman does not correspond to the Ediacaran-Cambrian boundary, but represents a facies change

Figure 29: *Archeocyathids from the Poleta formation in the Death Valley area*

from marine to evaporite-dominated strata — which would mean that dates from other, more suitable sections, ranging from 544 or 542 Ma, are more suitable.

Paleogeography

Plate reconstructions suggest a global supercontinent, Pannotia, was in the process of breaking up early in the period, with Laurentia (North America), Baltica, and Siberia having separated from the main supercontinent of Gondwana to form isolated land masses. Most continental land was clustered in the Southern Hemisphere at this time, but was drifting north. Large, high-velocity rotational movement of Gondwana appears to have occurred in the Early Cambrian.

With a lack of sea ice – the great glaciers of the Marinoan Snowball Earth were long melted – the sea level was high, which led to large areas of the continents being flooded in warm, shallow seas ideal for sea life. The sea levels fluctuated somewhat, suggesting there were 'ice ages', associated with pulses of expansion and contraction of a south polar ice cap.

In Baltoscandia a Lower Cambrian transgression transformed large swathes of the Sub-Cambrian peneplain into an epicontinental sea.

Climate

The Earth was generally cold during the early Cambrian, probably due to the ancient continent of Gondwana covering the South Pole and cutting off polar ocean currents. However, average temperatures were 7 degrees Celsius higher than today. There were likely polar ice caps and a series of glaciations, as the planet was still recovering from an earlier Snowball Earth. It became warmer towards the end of the period; the glaciers receded and eventually disappeared, and sea levels rose dramatically. This trend would continue into the Ordovician period.

Flora

Although there were a variety of macroscopic marine plantsWikipedia:Avoid weasel wordsWikipedia:Citation needed no land plant (embryophyte) fossils are known from the Cambrian. However, biofilms and microbial mats were well developed on Cambrian tidal flats and beaches 500 mya.,[116] and microbes forming microbial Earth ecosystems, comparable with modern soil crust of desert regions, contributing to soil formation.

Oceanic life

Life timeline

Axis scale: million years

Also see: *Human timeline* and *Nature timeline*

Most animal life during the Cambrian was aquatic. Trilobites were once assumed to be the dominant life form at that time, but this has proven to be incorrect. Arthropods were by far the most dominant animals in the ocean, but trilobites were only a minor part of the total arthropod diversity. What made them so apparently abundant was their heavy armor reinforced by calcium carbonate ($CaCO_3$), which fossilized far more easily than the fragile chitinous exoskeletons of other arthropods, leaving numerous preserved remains.

The period marked a steep change in the diversity and composition of Earth's biosphere. The Ediacaran biota suffered a mass extinction at the start of the Cambrian Period, which corresponded to an increase in the abundance and complexity of burrowing behaviour. This behaviour had a profound and irreversible effect on the substrate which transformed the seabed ecosystems. Before the Cambrian, the sea floor was covered by microbial mats. By the end of the Cambrian, burrowing animals had destroyed the mats in many areas through bioturbation, and gradually turned the seabeds into what they are today.Wikipedia:Please clarify As a consequence, many of those organisms that were dependent on the mats became extinct, while the other species adapted to the changed environment that now offered new ecological niches. Around the same time there was a seemingly rapid appearance of representatives of all the mineralized phyla except the Bryozoa, which appeared in the Lower Ordovician. However, many of those phyla were represented only by stem-group forms; and since mineralized phyla generally have a benthic origin, they may not be a good proxy for (more abundant) non-mineralized phyla.

While the early Cambrian showed such diversification that it has been named the Cambrian Explosion, this changed later in the period, when there occurred a sharp drop in biodiversity. About 515 million years ago, the number of species going extinct exceeded the number of new species appearing. Five million years later, the number of genera had dropped from an earlier peak of about 600 to just 450. Also, the speciation rate in many groups was reduced to between a fifth and a third of previous levels. 500 million years ago, oxygen levels fell dramatically in the oceans, leading to hypoxia, while the level of poisonous hydrogen sulfide simultaneously increased, causing another extinction. The later half of Cambrian was surprisingly barren and show evidence of several rapid extinction events; the stromatolites which had been replaced by reef building sponges known as Archaeocyatha, returned once more as the archaeocyathids became extinct. This declining trend did not change until the Great Ordovician Biodiversification Event.[117]

Figure 30: *A reconstruction of Margaretia dorus from the Burgess Shale, which were once believed to be green algae, but are now understood to represent hemichordates.*

Some Cambrian organisms ventured onto land, producing the trace fossils *Protichnites* and *Climactichnites*. Fossil evidence suggests that euthycarcinoids, an extinct group of arthropods, produced at least some of the *Protichnites*.[118,119] Fossils of the track-maker of *Climactichnites* have not been found; however, fossil trackways and resting traces suggest a large, slug-like mollusc.[120,121]

In contrast to later periods, the Cambrian fauna was somewhat restricted; free-floating organisms were rare, with the majority living on or close to the sea floor; and mineralizing animals were rarer than in future periods, in part due to the unfavourable ocean chemistry.

Many modes of preservation are unique to the Cambrian, and some preserve soft body parts, resulting in an abundance of *Lagerstätten*.

Symbol

The United States Federal Geographic Data Committee uses a "barred capital C" ☐☐☐ character to represent the Cambrian Period. The Unicode character is <templatestyles src="Mono/styles.css" />U+A792 ☐ LATIN CAPITAL LETTER C WITH BAR.[122]

Gallery

Figure 31: *Stromatolites of the Pika Formation (Middle Cambrian) near Helen Lake, Banff National Park, Canada*

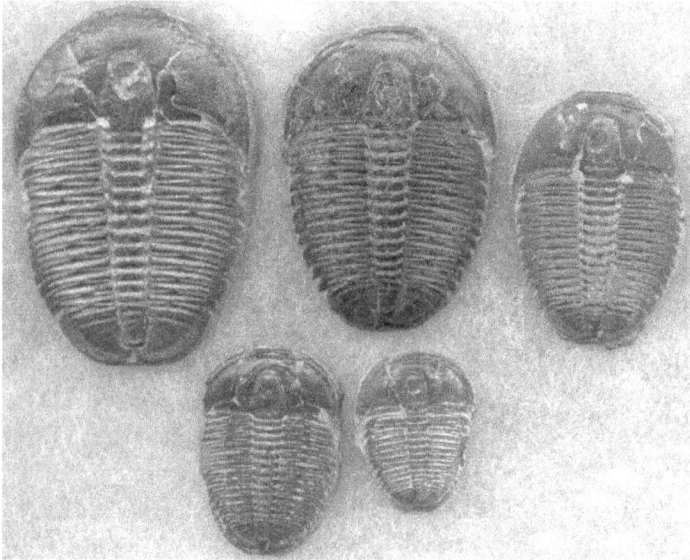

Figure 32: *Trilobites were very common during this time*

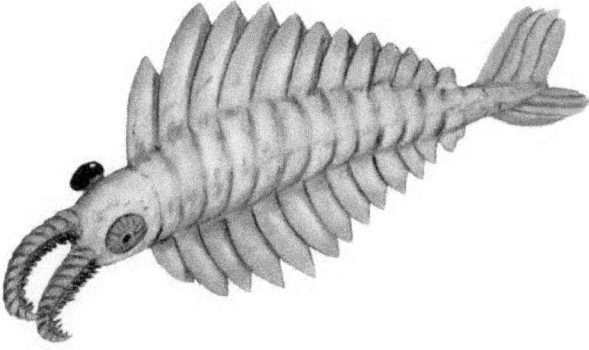

Figure 33: *Anomalocaris was an early marine predator, among the various arthropods of the time.*

Figure 34: *Pikaia was an early chordate from the Middle Cambrian*

Figure 35: *Opabinia was a creature with an unusual body plan; it was probably related to arthropods*

Figure 36: *Protichnites were the trackways of arthropods that walked Cambrian beaches.*

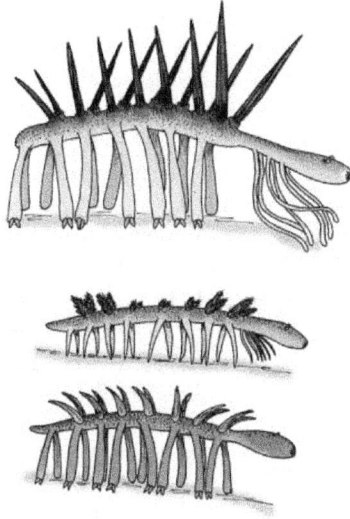

Figure 37: *Hallucigenia is maybe an early ancestor of the Velvet worms. Reconstructions of H. sparsa, H. hongmeia, and H. fortis*

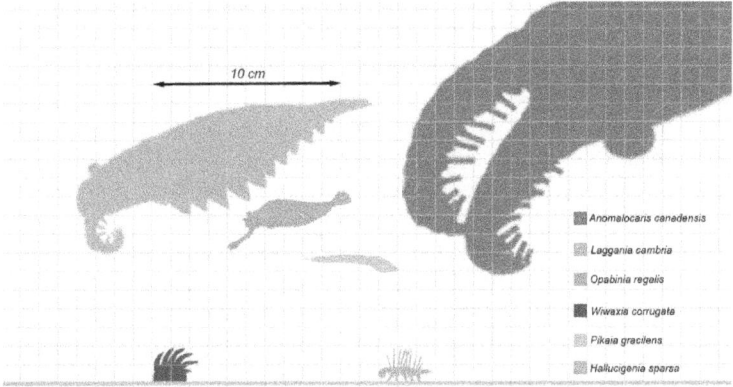

Anomalocaris canadensis
Laggania cambria
Opabinia regalis
Wiwaxia corrugata
Pikaia gracilens
Hallucigenia sparsa

Figure 38: *Size comparison of different Cambrian species*

Further reading

Wikisource has original works on the topic: *Paleozoic#Cambrian*

- Amthor, J. E.; Grotzinger, John P.; Schröder, Stefan; Bowring, Samuel A.; Ramezani, Jahandar; Martin, Mark W.; Matter, Albert (2003). "Extinction of *Cloudina* and *Namacalathus* at the Precambrian-Cambrian boundary in Oman". *Geology*. **31** (5): 431–434. Bibcode: 2003Geo....31..431A[123]. doi: 10.1130/0091-7613(2003)031<0431:EOCANA>2.0.CO;2[124].
- Collette, J. H.; Gass, K. C.; Hagadorn, J. W. (2012). "*Protichnites eremita* unshelled? Experimental model-based neoichnology and new evidence for a euthycarcinoid affinity for this ichnospecies". *Journal of Paleontology*. **86** (3): 442–454. doi: 10.1666/11-056.1[125].
- Collette, J. H.; Hagadorn, J. W. (2010). "Three-dimensionally preserved arthropods from Cambrian Lagerstatten of Quebec and Wisconsin". *Journal of Paleontology*. **84** (4): 646–667. doi: 10.1666/09-075.1[126].
- Getty, P. R.; Hagadorn, J. W. (2008). "Reinterpretation of *Climactichnites* Logan 1860 to include subsurface burrows, and erection of *Musculopodus* for resting traces of the trailmaker". *Journal of Paleontology*. **82** (6): 1161–1172. doi: 10.1666/08-004.1[127].
- Gould, S. J.; Wonderful Life: the Burgess Shale and the Nature of Life (New York: Norton, 1989)
- Ogg, J.; June 2004, *Overview of Global Boundary Stratotype Sections and Points (GSSPs)* https://web.archive.org/web/20060423084018/http://www.stratigraphy.org/gssp.htm Accessed 30 April 2006.
- Owen, R. (1852). "Description of the impressions and footprints of the *Protichnites* from the Potsdam sandstone of Canada". *Geological Society of London Quarterly Journal*. **8**: 214–225. doi: 10.1144/GSL.JGS.1852.008.01-02.26[128].
- Peng, S.; Babcock, L.E.; Cooper, R.A. (2012). "The Cambrian Period". *The Geologic Time Scale*[129] (PDF).
- Schieber, J.; Bose, P. K.; Eriksson, P. G.; Banerjee, S.; Sarkar, S.; Altermann, W.; Catuneau, O. (2007). *Atlas of Microbial Mat Features Preserved within the Clastic Rock Record*. Elsevier. pp. 53–71.
- Yochelson, E. L.; Fedonkin, M. A. (1993). "Paleobiology of *Climactichnites*, and Enigmatic Late Cambrian Fossil"[130] (Free full text). *Smithsonian Contributions to Paleobiology*. **74** (74): 1–74. doi: 10.5479/si.00810266.74.1[131].

External links

> Wikimedia Commons has media related to *Cambrian*.

- Cambrian period[132] on *In Our Time* at the BBC
- Biostratigraphy[133] – includes information on Cambrian trilobite biostratigraphy
- Dr. Sam Gon's trilobite pages[134] (contains numerous Cambrian trilobites)
- Examples of Cambrian Fossils[135]
- Paleomap Project[136]
- Report on the web on Amthor and others from *Geology* vol. 31[137]
- Weird Life on the Mats[138]

Ordovician

Ordovician Period _485.4–443.8 million years ago_ PreЄЄ OSD C P T J K PgN	
Mean atmospheric O $_2$ content over period duration	c. 13.5 vol %[139,140] (68 % of modern level
Mean atmospheric CO $_2$ content over period duration	c. 4200 ppm[141] (15 times pre-industrial level)
Mean surface tempera- ture over period duration	c. 16 °C[142] (2 °C above modern level)
Sea level (above present day)	180 m; rising to 220 m in Caradoc and falling sharply to 140 m in end-Ordovician glaciations

Key events in the Ordovician

Key events of the Ordovician Period.
ICS approved stages.
Axis scale: millions of years ago.

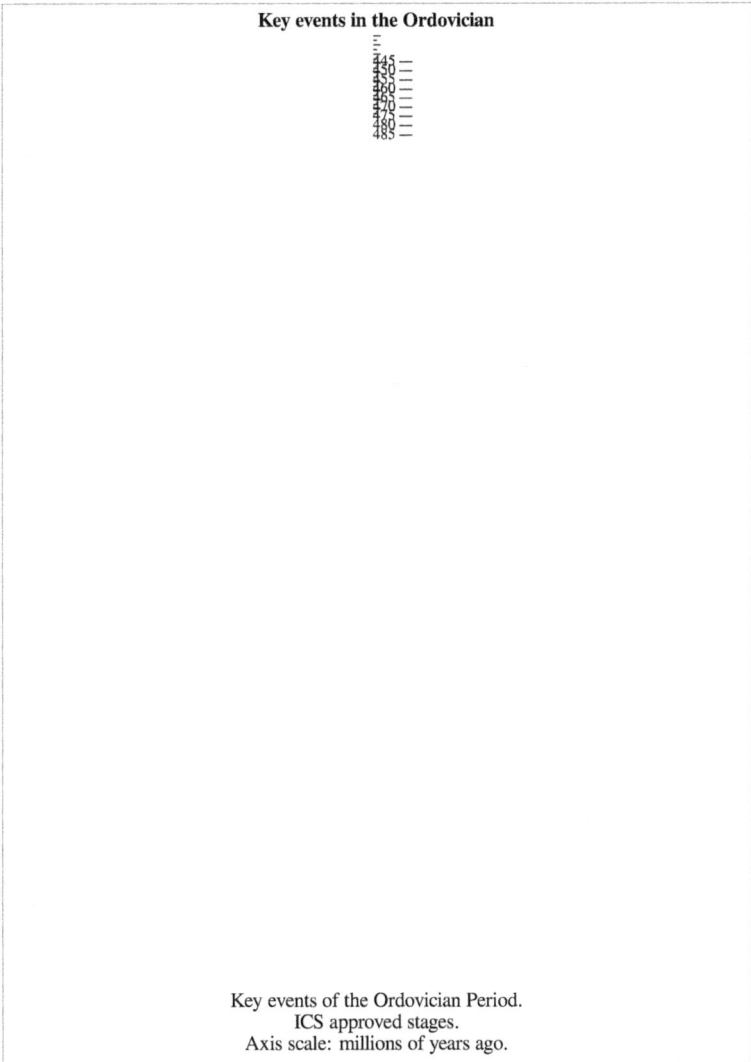

The **Ordovician** (/ɔːrdəˈvɪʃən/) is a geologic period and system, the second of six periods of the Paleozoic Era. The Ordovician spans 41.2 million years from the end of the Cambrian Period 485.4 million years ago (Mya) to the start of the Silurian Period 443.8 Mya.

The Ordovician, named after the Celtic tribe of the Ordovices, was defined by Charles Lapworth in 1879 to resolve a dispute between followers of Adam Sedgwick and Roderick Murchison, who were placing the same rock beds in northern Wales into the Cambrian and Silurian systems, respectively.[143] Lapworth recognized that the fossil fauna in the disputed strata were different from those of either the Cambrian or the Silurian systems, and placed them in a system of their own. The Ordovician received international approval in 1960 (forty years after Lapworth's death), when it was adopted as an official period of the Paleozoic Era by the International Geological Congress.

Life continued to flourish during the Ordovician as it did in the earlier Cambrian period, although the end of the period was marked by the Ordovician–Silurian extinction event. Invertebrates, namely molluscs and arthropods, dominated the oceans. The Great Ordovician Biodiversification Event considerably increased the diversity of life. Fish, the world's first true vertebrates, continued to evolve, and those with jaws may have first appeared late in the period. Life had yet to diversify on land. About 100 times as many meteorites struck the Earth per year during the Ordovician compared with today.

Dating: extinction events

The Ordovician Period began with a major extinction called the Cambrian–Ordovician extinction event, about 485.4 Mya (million years ago). It lasted for about 42 million years and ended with the Ordovician–Silurian extinction event, about 443.8 Mya (ICS, 2004) which wiped out 60% of marine genera. The dates given are recent radiometric dates and vary slightly from those found in other sources. This second period of the Paleozoic era created abundant fossils that became major petroleum and gas reservoirs.

The boundary chosen for the beginning of both the Ordovician Period and the Tremadocian stage is highly significant. It correlates well with the occurrence of widespread graptolite, conodont, and trilobite species. The base (start) of the Tremadocian allows scientists to relate these species not only to each other, but also to species that occur with them in other areas. This makes it easier to place many more species in time relative to the beginning of the Ordovician Period.

Subdivisions

A number of regional terms have been used to subdivide the Ordovician Period. In 2008, the ICS erected a formal international system of subdivisions, illustrated to the right.[144] There exist Baltoscandic, British, Siberian, North American, Australian, Chinese Mediterranean&North-Gondwanan regional stratigraphic schemes.[145]

Ordoviciani regional stages

Epoch	Stage	British epoch	British stage	North American epoch	North American stage	Australian epoch	Australian stage	Chinese epoch	Chinese stage
Late Ordovician	Hirnantian stage	Ashgill stage	Hirnant stage	Cincinnati series	Gamach stage	Late Ordovician	Bolinda stage	Late Ordovician	Hirnant stage
	Katy stage		Rawthey stage		Richmond stage				Chientangkiang stage
			Cautley stage		Maysville stage		Easton stage		Neichiashan stage
			Pusgill stage		Eden stage				
		Caradoc series	Strefford stage	Mohawk stage	Chatfield stage				
			Cheney stage						
	Sandby stage		Burrell stage		Turin stage		Gisborne stage		
			Aureluc stage	Whiterock stage	Chazy stage				
Middle Ordovician	Darriwil stage	Llanvirn series	Llandeil stage			Middle Ordovician	Darriwil stage	Middle Ordovician	Darriwil stage
			Abereiddy stage		Not defined				
	Daping stage	Arenig series	Fenn stage			Early Ordovician	Yapeen stage		Daping stage
			Whitland stage		Ranger stage		Castlemaine stage		
				Ibex series	Black Hills stage		Chewton stage		
							Bendigo stage		
Early Ordovician	Flo stage		Moridun stage		Tule stage		Lancefield stage	Early Ordovician	Flo stage
	Tremadoc stage	Tremadoc series	Migneint stage		Stairs stage				Tremadoc stage
			Cressage stage		Skullrock stage				

The Ordovician Period in Britain was traditionally broken into Early (Tremadocian and Arenig), Middle (Llanvirn (subdivided into Abereiddian and Llandeilian) and Llandeilo) and Late (Caradoc and Ashgill) epochs. The corresponding rocks of the Ordovician System are referred to as coming from the Lower, Middle, or Upper part of the column. The faunal stages (subdivisions of epochs) from youngest to oldest are:

Late Ordovician

- Hirnantian/Gamach (Ashgill)
- Rawtheyan/Richmond (Ashgill)
- Cautleyan/Richmond (Ashgill)
- Pusgillian/Maysville/Richmond (Ashgill)

Middle Ordovician

- Trenton (Caradoc)
- Onnian/Maysville/Eden (Caradoc)
- Actonian/Eden (Caradoc)
- Marshbrookian/Sherman (Caradoc)
- Longvillian/Sherman (Caradoc)
- Soudleyan/Kirkfield (Caradoc)
- Harnagian/Rockland (Caradoc)
- Costonian/Black River (Caradoc)
- Chazy (Llandeilo)
- Llandeilo (Llandeilo)
- Whiterock (Llanvirn)
- Llanvirn (Llanvirn)

Early Ordovician

- Cassinian (Arenig)
- Arenig/Jefferson/Castleman (Arenig)
- Tremadoc/Deming/Gaconadian (Tremadoc)

British stages

The Tremadoc corresponds to the (modern) Tremadocian. The Floian corresponds to the lower Arenig; the Arenig continues until the early Darriwilian, subsuming the Dapingian. The Llanvirn occupies the rest of the Darriwilian, and terminates with it at the base of the Late Ordovician. The Sandbian represents the first half of the Caradoc; the Caradoc ends in the mid-Katian, and the Ashgill represents the last half of the Katian, plus the Hirnantian.Wikipedia:Cleanup

Paleogeography

During the Ordovician, the southern continents were collected into Gondwana. Gondwana started the period in equatorial latitudes and, as the period progressed, drifted toward the South Pole.

Early in the Ordovician, the continents of Laurentia (in present-day North America), Siberia, and Baltica (present-day northern Europe) were still independent continents (since the break-up of the supercontinent Pannotia earlier), but Baltica began to move towards Laurentia later in the period, causing the Iapetus Ocean between them to shrink. The small continent Avalonia separated from Gondwana and began to move north towards Baltica and Laurentia, opening the Rheic Ocean between Gondwana and Avalonia.

The Taconic orogeny, a major mountain-building episode, was well under way in Cambrian times. In the early and middle Ordovician, temperatures were mild, but at the beginning of the Late Ordovician, from 460 to 450 Ma, volcanoes along the margin of the Iapetus Ocean spewed massive amounts of carbon dioxide, a greenhouse gas, into the atmosphere, turning the planet into a hothouse.

Initially, sea levels were high, but as Gondwana moved south, ice accumulated into glaciers and sea levels dropped. At first, low-lying sea beds increased diversity, but later glaciation led to mass extinctions as the seas drained and continental shelves became dry land. During the Ordovician, in fact during the Tremadocian, marine transgressions worldwide were the greatest for which evidence is preserved.

These volcanic island arcs eventually collided with proto North America to form the Appalachian mountains. By the end of the Late Ordovician the volcanic emissions had stopped. Gondwana had by that time neared the south pole and was largely glaciated.

Ordovician meteor event

The Ordovician meteor event is a proposed shower of meteors that occurred during the Middle Ordovician period, roughly 470 million years ago. It is not associated with any major extinction event.

Figure 39: *External mold of Ordovician bivalve showing that the original aragonite shell dissolved on the sea floor, leaving a cemented mold for biological encrustation (Waynesville Formation of Franklin County, Indiana).*

Geochemistry

The Ordovician was a time of calcite sea geochemistry in which low-magnesium calcite was the primary inorganic marine precipitate of calcium carbonate. Carbonate hardgrounds were thus very common, along with calcitic ooids, calcitic cements, and invertebrate faunas with dominantly calcitic skeletons. Biogenic aragonite, like that composing the shells of most mollusks, dissolved rapidly on the sea floor after death.

Unlike Cambrian times, when calcite production was dominated by microbial and non-biological processes, animals (and macroalgae) became a dominant source of calcareous material in Ordovician deposits.

Climate and sea level

The Ordovician saw the highest sea levels of the Paleozoic, and the low relief of the continents led to many shelf deposits being formed under hundreds of metres of water. The sea level rose more or less continuously throughout the Early Ordovician, leveling off somewhat during the middle of the period. Locally, some regressions occurred, but sea level rise continued in the beginning of the Late Ordovician. Sea levels fell steadily in accord with the cooling temperatures for ~30 million years leading up to the Hirnantian glaciation. During this icy stage, sea level seems to have risen and dropped somewhat, but despite much study the details remain unresolved.

At the beginning of the period, around 485.4 million years ago, the climate was very hot due to high concentration of CO_2 (4200 ppm) in the atmosphere, which gave a strong greenhouse effect. By contrast, today the concentration is just above 400 ppm. Marine water temperatures are assumed to have averaged 45 °C (113 °F), which restricted the diversification of complex multi-cellular organisms. But over time, the climate became cooler, and around 460 million years ago, the ocean temperatures became comparable to those of present-day equatorial waters.[146]

As with North America and Europe, Gondwana was largely covered with shallow seas during the Ordovician. Shallow clear waters over continental shelves encouraged the growth of organisms that deposit calcium carbonates in their shells and hard parts. The Panthalassic Ocean covered much of the northern hemisphere, and other minor oceans included Proto-Tethys, Paleo-Tethys, Khanty Ocean, which was closed off by the Late Ordovician, Iapetus Ocean, and the new Rheic Ocean.

As the Ordovician progressed, we see evidence of glaciers on the land we now know as Africa and South America, which were near the South Pole at the time, and covered by ice caps.

Life

For most of the Late Ordovician life continued to flourish, but at and near the end of the period there were mass-extinction events that seriously affected planktonic forms like conodonts and graptolites. The trilobites Agnostida and Ptychopariida completely died out, and the Asaphida were much reduced. Brachiopods, bryozoans and echinoderms were also heavily affected, and the endocerid cephalopods died out completely, except for possible rare Silurian forms. The Ordovician–Silurian Extinction Events may have been caused by an ice age that occurred at the end of the Ordovician period, due to the expansion of the first terrestrial plants,[147] as the end of the Late Ordovician was one of the coldest times in the last 600 million years of earth history.

Fauna

On the whole, the fauna that emerged in the Ordovician were the template for the remainder of the Palaeozoic. The fauna was dominated by tiered communities of suspension feeders, mainly with short food chains. The ecological

Figure 40: *A diorama depicting Ordovician flora and fauna.*

Figure 41: *Nautiloids like Orthoceras were among the largest predators in the Ordovician.*

Figure 42: *Fossiliferous limestone slab from the Liberty Formation (Upper Ordovician) of Caesar Creek State Park near Waynesville, Ohio.*

system reached a new grade of complexity far beyond that of the Cambrian fauna, which has persisted until the present day.

Though less famous than the Cambrian explosion, the Ordovician radiation was no less remarkable; marine faunal genera increased fourfold, resulting in 12% of all known Phanerozoic marine fauna. Another change in the fauna was the strong increase in filter-feeding organisms.[148] The trilobite, inarticulate brachiopod, archaeocyathid, and eocrinoid faunas of the Cambrian were succeeded by those that dominated the rest of the Paleozoic, such as articulate brachiopods, cephalopods, and crinoids. Articulate brachiopods, in particular, largely replaced trilobites in shelf communities. Their success epitomizes the greatly increased diversity of carbonate shell-secreting organisms in the Ordovician compared to the Cambrian.

In North America and Europe, the Ordovician was a time of shallow continental seas rich in life. Trilobites and brachiopods in particular were rich and diverse. Although solitary corals date back to at least the Cambrian, reef-forming corals appeared in the early Ordovician, corresponding to an increase in the stability of carbonate and thus a new abundance of calcifying animals.

Molluscs, which appeared during the Cambrian or even the Ediacaran, became common and varied, especially bivalves, gastropods, and nautiloid
cephalopods.

Now-extinct marine animals called graptolites thrived in the oceans. Some
new cystoids and crinoids appeared.

It was long thought that the first true vertebrates (fish — Ostracoderms) appeared in the Ordovician, but recent discoveries in China reveal that they probably originated in the Early Cambrian.Wikipedia:Citation needed The very
first gnathostome (jawed fish) appeared in the Late Ordovician epoch.

During the Middle Ordovician there was a large increase in the intensity and
diversity of bioeroding organisms. This is known as the Ordovician Bioerosion
Revolution. It is marked by a sudden abundance of hard substrate trace fossils such as *Trypanites*, *Palaeosabella*, *Petroxestes* and *Osprioneides*. Several
groups of endobiotic symbionts appeared in the Ordovician.

In the Early Ordovician, trilobites were joined by many new types of organisms, including tabulate corals, strophomenid, rhynchonellid, and many
new orthid brachiopods, bryozoans, planktonic graptolites and conodonts, and
many types of molluscs and echinoderms, including the ophiuroids ("brittle
stars") and the first sea stars. Nevertheless, the trilobites remained abundant,
all the Late Cambrian orders continued, and were joined by the new group
Phacopida. The first evidence of land plants also appeared; see Evolutionary
history of life.

In the Middle Ordovician, the trilobite-dominated Early Ordovician communities were replaced by generally more mixed ecosystems, in which brachiopods,
bryozoans, molluscs, cornulitids, tentaculitids and echinoderms all flourished,
tabulate corals diversified and the first rugose corals appeared. The planktonic graptolites remained diverse, with the Diplograptina making their appearance. Bioerosion became an important process, particularly in the thick
calcitic skeletons of corals, bryozoans and brachiopods, and on the extensive
carbonate hardgrounds that appear in abundance at this time. One of the earliest known armoured agnathan ("ostracoderm") vertebrate, *Arandaspis*, dates
from the Middle Ordovician.

Trilobites in the Ordovician were very different from their predecessors in
the Cambrian. Many trilobites developed bizarre spines and nodules to de-
fend against predators such as primitive eurypterids and nautiloids while other
trilobites such as *Aeglina prisca* evolved to become swimming forms. Some
trilobites even developed shovel-like snouts for ploughing through muddy sea
bottoms. Another unusual clade of trilobites known as the trinucleids developed a broad pitted margin around their head shields. Some trilobites such
as *Asaphus kowalewski* evolved long eyestalks to assist in detecting predators

whereas other trilobite eyes in contrast disappeared completely.[149] Molecular clock analyses suggest that early arachnids started living on land by the end of the Ordovician.

The earliest known octocorals date from the Ordovician.

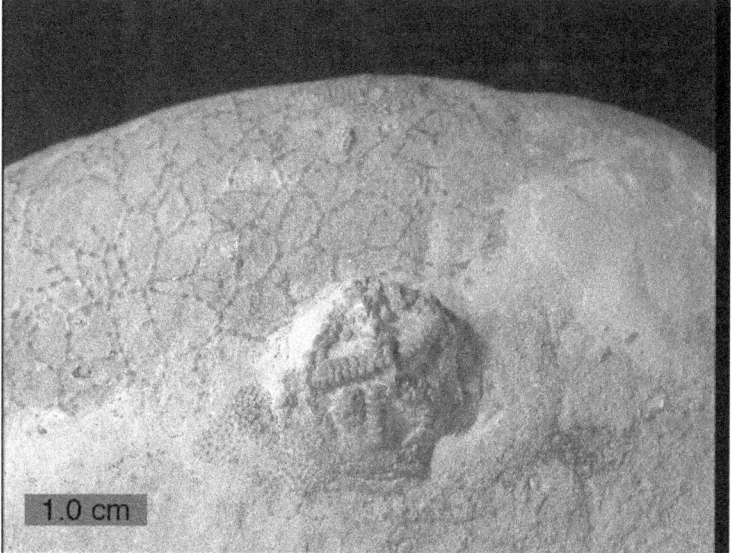

Figure 43: *The Upper Ordovician edrioasteroid Cystaster stellatus on a cobble from the Kope Formation in northern Kentucky. In the background is the cyclostome bryozoan Corynotrypa.*

Figure 44: *Fossil Mountain, west-central Utah; Middle Ordovician fossiliferous shales and limestones in the lower half.*

Figure 45: *Outcrop of Upper Ordovician rubbly limestone and shale, southern Indiana; College of Wooster students.*

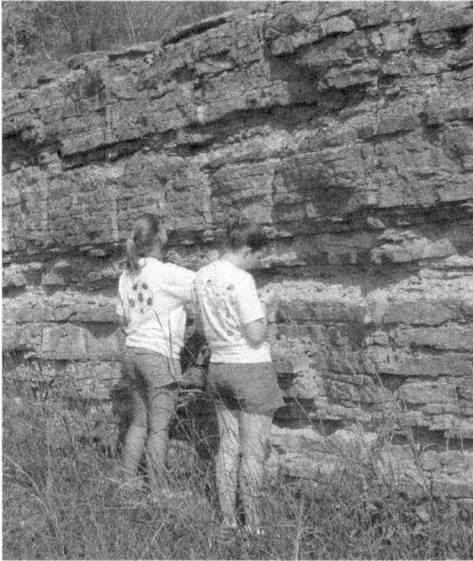

Figure 46: *Outcrop of Upper Ordovician limestone and minor shale, central Tennessee; College of Wooster students.*

Figure 47: *Trypanites borings in an Ordovician hardground, southeastern Indiana.*[150]

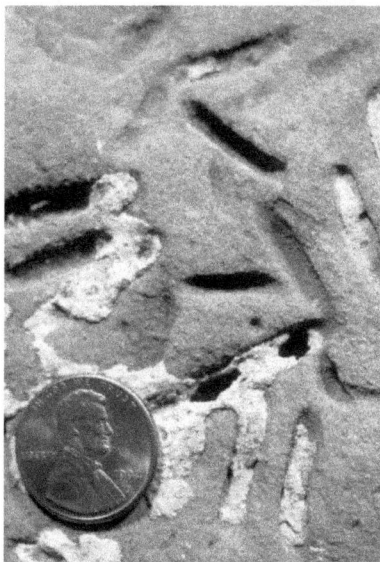

Figure 48: *Petroxestes borings in an Ordovician hardground, southern Ohio.*[151]

Figure 49: *Outcrop of Ordovician kukersite oil shale, northern Estonia.*

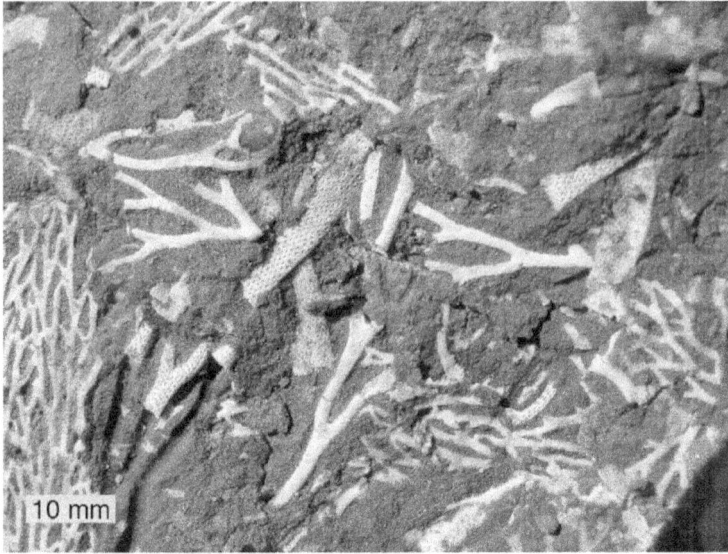

Figure 50: *Bryozoan fossils in Ordovician kukersite oil shale, northern Estonia.*

Figure 51: *Brachiopods and bryozoans in an Ordovician limestone, southern Minnesota.*

Figure 52: *Vinlandostrophia ponderosa, Maysvillian (Upper Ordovician) near Madison, Indiana. Scale bar is 5.0 mm.*

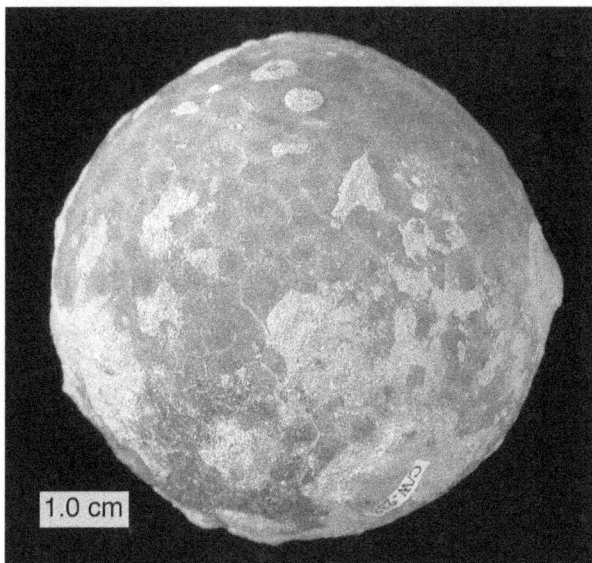

Figure 53: *The Ordovician cystoid Echinosphaerites (an extinct echinoderm) from northeastern Estonia; approximately 5 cm in diameter.*

Figure 54: *Prasopora, a trepostome bryozoan from the Ordovician of Iowa.*

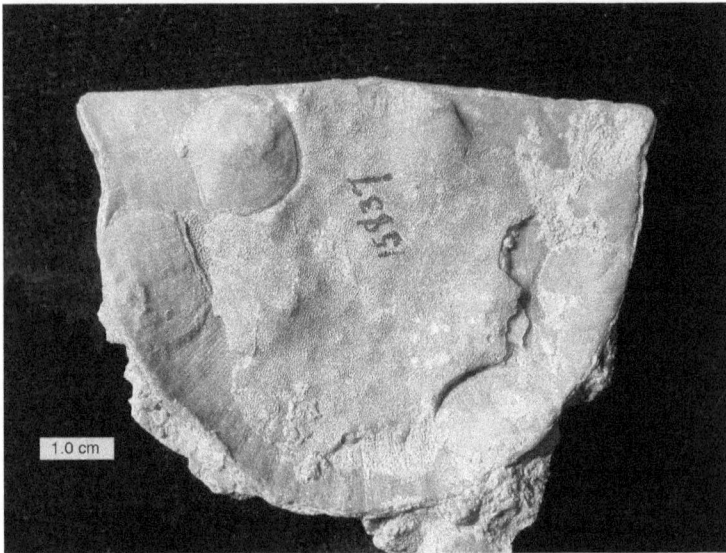

Figure 55: *An Ordovician strophomenid brachiopod*
with encrusting inarticulate brachiopods and a bryozoan.

Figure 56: *The heliolitid coral Protaraea richmondensis encrusting a gastropod; Cincinnatian (Upper Ordovician) of southeastern Indiana.*

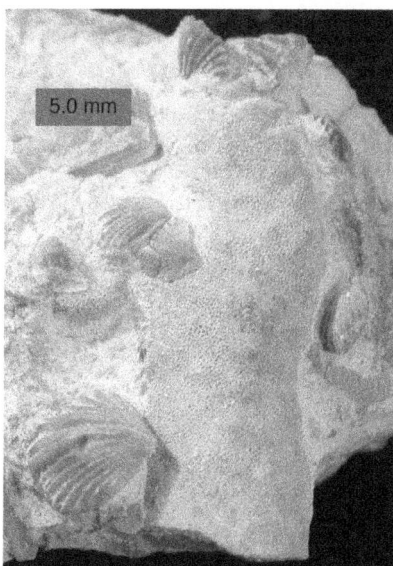

Figure 57: *Zygospira modesta, atrypid brachiopods, preserved in their original positions on a trepostome bryozoan; Cincinnatian (Upper Ordovician) of southeastern Indiana.*

Figure 58: *Graptolites (Amplexograptus) from the Ordovician near Caney Springs, Tennessee.*

Flora

Green algae were common in the Late Cambrian (perhaps earlier) and in the Ordovician. Terrestrial plants probably evolved from green algae, first appearing as tiny non-vascular forms resembling liverworts. Fossil spores from land plants have been identified in uppermost Ordovician sediments.

Among the first land fungi may have been arbuscular mycorrhiza fungi (Glomerales), playing a crucial role in facilitating the colonization of land by plants through mycorrhizal symbiosis, which makes mineral nutrients available to plant cells; such fossilized fungal hyphae and spores from the Ordovician of Wisconsin have been found with an age of about 460 million years ago, a time when the land flora most likely only consisted of plants similar to non-vascular bryophytes.

End of the period

The Ordovician came to a close in a series of extinction events that, taken together, comprise the second largest of the five major extinction events in Earth's history in terms of percentage of genera that became extinct. The only larger one was the Permian-Triassic extinction event.

Figure 59: *Colonization of land would have been limited to shorelines*

The extinctions occurred approximately 447–444 million years ago and mark the boundary between the Ordovician and the following Silurian Period. At that time all complex multicellular organisms lived in the sea, and about 49% of genera of fauna disappeared forever; brachiopods and bryozoans were greatly reduced, along with many trilobite, conodont and graptolite families.

The most commonly accepted theory is that these events were triggered by the onset of cold conditions in the late Katian, followed by an ice age, in the Hirnantian faunal stage, that ended the long, stable greenhouse conditions typical of the Ordovician.

The ice age was possibly not long-lasting. Oxygen isotopes in fossil brachiopods show its duration may have been only 0.5 to 1.5 million years. Other researchers (Page et al.) estimate more temperate conditions did not return until the late Silurian.

The late Ordovician glaciation event was preceded by a fall in atmospheric carbon dioxide (from 7000 ppm to 4400 ppm).[152] The dip was triggered by a burst of volcanic activity that deposited new silicate rocks, which draw CO_2 out of the air as they erode. This selectively affected the shallow seas where most organisms lived. As the southern supercontinent Gondwana drifted over the South Pole, ice caps formed on it, which have been detected in Upper

Ordovician rock strata of North Africa and then-adjacent northeastern South America, which were south-polar locations at the time.

As glaciers grew, the sea level dropped, and the vast shallow intra-continental Ordovician seas withdrew, which eliminated many ecological niches. When they returned, they carried diminished founder populations that lacked many whole families of organisms. They then withdrew again with the next pulse of glaciation, eliminating biological diversity with each change.[153] Species limited to a single epicontinental sea on a given landmass were severely affected. Tropical lifeforms were hit particularly hard in the first wave of extinction, while cool-water species were hit worst in the second pulse.

Those species able to adapt to the changing conditions survived to fill the ecological niches left by the extinctions.

At the end of the second event, melting glaciers caused the sea level to rise and stabilise once more. The rebound of life's diversity with the permanent re-flooding of continental shelves at the onset of the Silurian saw increased biodiversity within the surviving Orders.

An alternate extinction hypothesis suggested that a ten-second gamma-ray burst could have destroyed the ozone layer and exposed terrestrial and marine surface-dwelling life to deadly ultraviolet radiation and initiated global cooling.

Recent work considering the sequence stratigraphy of the Late Ordovician argues that the mass extinction was a single protracted episode lasting several hundred thousand years, with abrupt changes in water depth and sedimentation rate producing two pulses of last occurrences of species[154].

External links

Wikisource has original works on the topic: *Paleozoic#Ordovician*

Wikimedia Commons has media related to *Ordovician*.

- Ogg, Jim (June 2004). "Overview of Global Boundary Stratotype Sections and Points (GSSP's)"[155]. Archived from the original[156] on 2006-04-23. Retrieved 2006-04-30.
- Mehrtens, Charlotte. "Chazy Reef at Isle La Motte"[157]. An Ordovician reef in Vermont.
- Ordovician fossils of the famous Cincinnatian Group[158]

- The Dry Dredgers, an active group of amateur paleontologists in the Cincinnati area[159]

Silurian

Silurian Period	
443.8–419.2 million years ago	
PreЄЄ OSD C P T J K PgN	
Mean atmospheric O$_2$ content over period duration	c. 14 vol %[160,161] (70 % of modern level
Mean atmospheric CO$_2$ content over period duration	c. 4500 ppm[162] (16 times pre-industrial level)
Mean surface temperature over period duration	c. 17 °C[163] (3 °C above modern level)
Sea level (above present day)	Around 180 m, with short-term negative excursions

Epochs in the Silurian

Epochs of the Silurian Period.
Axis scale: millions of years ago.

The **Silurian** is a geologic period and system spanning 24.6 million years from the end of the Ordovician Period, at 443.8 million years ago (Mya), to the beginning of the Devonian Period, 419.2 Mya. As with other geologic periods, the rock beds that define the period's start and end are well identified, but the exact dates are uncertain by several million years. The base of the Silurian is set at a series of major Ordovician–Silurian extinction events when 60% of marine species were wiped out.

A significant evolutionary milestone during the Silurian was the diversification of jawed fish and bony fish. Multi-cellular life also began to appear on land in the form of small, bryophyte-like and vascular plants that grew beside lakes, streams, and coastlines, and terrestrial arthropods are also first found on land during the Silurian. However, terrestrial life would not greatly diversify and affect the landscape until the Devonian.

History of study

Life timeline

θ —

500 —
1000 —
1500 —
2000 —
2500 —
3000 —
3500 —
4000 —
4500 —

Axis scale: million years

Also see: *Human timeline* and *Nature timeline*

The Silurian system was first identified by British geologist Roderick Murchison, who was examining fossil-bearing sedimentary rock strata in south Wales in the early 1830s. He named the sequences for a Celtic tribe of Wales, the Silures, inspired by his friend Adam Sedgwick, who had named the period of his study the Cambrian, from the Latin name for Wales. This naming does not indicate any correlation between the occurrence of the Silurian rocks and the land inhabited by the Silures (cf. Geologic map of Wales, Map of pre-Roman tribes of Wales). In 1835 the two men presented a joint paper, under the title *On the Silurian and Cambrian Systems, Exhibiting the Order in which the Older Sedimentary Strata Succeed each other in England and Wales,* which was the germ of the modern geological time scale. As it was first identified, the "Silurian" series when traced farther afield quickly came to overlap Sedgwick's "Cambrian" sequence, however, provoking furious disagreements that ended the friendship. Charles Lapworth resolved the conflict by defining a new Ordovician system including the contested beds. An early alternative name for the Silurian was *"Gotlandian"* after the strata of the Baltic island of Gotland.

The French geologist Joachim Barrande, building on Murchison's work, used the term *Silurian* in a more comprehensive sense than was justified by subsequent knowledge. He divided the Silurian rocks of Bohemia into eight stages. His interpretation was questioned in 1854 by Edward Forbes, and the later stages of Barrande, F, G and H, have since been shown to be Devonian. Despite these modifications in the original groupings of the strata, it is recognized that Barrande established Bohemia as a classic ground for the study of the earliest fossils.

Subdivisions

Llandovery

The Llandovery Epoch lasted from $443.8^{164} \pm 1.5$ to $433.4^{165} \pm 2.8$ mya, and is subdivided into three stages: the Rhuddanian,[166] lasting until 440.8^{167} million years ago, the Aeronian, lasting to 438.5^{168} million years ago, and the Telychian. The epoch is named for the town of Llandovery in Carmarthenshire, Wales.

Wenlock

The Wenlock, which lasted from $433.4^{165} \pm 1.5$ to $427.4^{169} \pm 2.8$ mya, is sub-divided into the Sheinwoodian (to 430.5^{170} million years ago) and Homerian ages. It is named after Wenlock Edge in Shropshire, England. During the Wenlock, the oldest-known tracheophytes of the genus *Cooksonia*, appear. The complexity of slightly later Gondwana plants like *Baragwanathia*, which resembled a modern clubmoss, indicates a much longer history for vascular plants, extending into the early Silurian or even Ordovician.Wikipedia:Citation needed The first terrestrial animals also appear in the Wenlock, represented by air-breathing millipedes from Scotland.

Ludlow

The Ludlow, lasting from $427.4^{169} \pm 1.5$ to $423^{171} \pm 2.8$ mya, comprises the Gorstian stage, lasting until 425.6^{172} million years ago, and the Ludfordian stage. It is named for the town of Ludlow (and neighbouring Ludford) in Shropshire, England.

Přídolí

The Přídolí, lasting from $423^{171} \pm 1.5$ to $419.2^{173} \pm 2.8$ mya, is the final and shortest epoch of the Silurian. It is named after one locality at the *Homolka a Přídolí* nature reserve near the Prague suburb Slivenec in the Czech Republic. *Přídolí* is the old name of a cadastral field area.

Regional stages

In North America a different suite of regional stages is sometimes used:

• Cayugan (Late Silurian – Ludlow)
• Lockportian (Middle Silurian: late Wenlock)
• Tonawandan (Middle Silurian: early Wenlock)
• Ontarian (Early Silurian: late Llandovery)
• Alexandrian (Earliest Silurian: early Llandovery)

In Estonia the following suite of regional stages is used:[174]

• Ohessaare stage (Late Silurian – early Přídolí)
• Kaugatuma stage (Late Silurian – late Přídolí)
• Kuressaare stage (Late Silurian – late Ludlow)
• Paadla stage (Late Silurian – early Ludlow)
• Rootsiküla stage (Middle Silurian: late Wenlock)
• Jaagarahu stage (Middle Silurian: middle Wenlock)
• Jaani stage (Middle Silurian: early Wenlock)
• Adavere stage (Early Silurian: late Llandovery)

Figure 60: *Ordovician-Silurian boundary on Hovedøya, Norway, show-ing brownish late Ordovician mudstone and later dark deep-water Sil-urian shale. The layers have been overturned by the Caledonian orogeny.*

- Raikküla stage (Early Silurian: middle Llandovery)
- Juuru stage (Earliest Silurian: early Llandovery)

Geography

With the supercontinent Gondwana covering the equator and much of the southern hemisphere, a large ocean occupied most of the northern half of the globe. The high sea levels of the Silurian and the relatively flat land (with few significant mountain belts) resulted in a number of island chains, and thus a rich diversity of environmental settings.

During the Silurian, Gondwana continued a slow southward drift to high south-ern latitudes, but there is evidence that the Silurian icecaps were less extensive than those of the late-Ordovician glaciation. The southern continents remained united during this period. The melting of icecaps and glaciers contributed to a rise in sea level, recognizable from the fact that Silurian sediments overlie eroded Ordovician sediments, forming an unconformity. The continents of Avalonia, Baltica, and Laurentia drifted together near the equator, starting the formation of a second supercontinent known as Euramerica.

When the proto-Europe collided with North America, the collision folded coastal sediments that had been accumulating since the Cambrian off the east coast of North America and the west coast of Europe. This event is the Caledonian orogeny, a spate of mountain building that stretched from New York State through conjoined Europe and Greenland to Norway. At the end of the Silurian, sea levels dropped again, leaving telltale basins of evaporites extending from Michigan to West Virginia, and the new mountain ranges were rapidly eroded. The Teays River, flowing into the shallow mid-continental sea, eroded Ordovician Period strata, forming deposits of Silurian strata in northern Ohio and Indiana.

The vast ocean of Panthalassa covered most of the northern hemisphere. Other minor oceans include two phases of the Tethys, the Proto-Tethys and Paleo-Tethys, the Rheic Ocean, the Iapetus Ocean (a narrow seaway between Avalonia and Laurentia), and the newly formed Ural Ocean.

Climate and sea level

The Silurian period enjoyed relatively stable and warm temperatures, in contrast with the extreme glaciations of the Ordovician before it, and the extreme heat of the ensuing Devonian. Sea levels rose from their Hirnantian low throughout the first half of the Silurian; they subsequently fell throughout the rest of the period, although smaller scale patterns are superimposed on this general trend; fifteen high-stands can be identified, and the highest Silurian sea level was probably around 140 m higher than the lowest level reached.

During this period, the Earth entered a long, warm greenhouse phase, supported by high CO_2 levels of 4500 ppm, and warm shallow seas covered much of the equatorial land masses. Early in the Silurian, glaciers retreated back into the South Pole until they almost disappeared in the middle of Silurian. The period witnessed a relative stabilization of the Earth's general climate, ending the previous pattern of erratic climatic fluctuations. Layers of broken shells (called coquina) provide strong evidence of a climate dominated by violent storms generated then as now by warm sea surfaces. Later in the Silurian, the climate cooled slightly, but closer to the Silurian-Devonian boundary, the climate became warmer.Wikipedia:Citation needed

Perturbations

The climate and carbon cycle appears to be rather unsettled during the Silurian, which has a higher concentration of isotopic excursions than any other period. The Ireviken event, Mulde event and Lau event each represent isotopic excursions following a minor mass extinction and associated with rapid sea-level change, in addition to the larger extinction at the end of the Silurian. Each one leaves a similar signature in the geological record, both geochemically and biologically; pelagic (free-swimming) organisms were particularly hard hit, as were brachiopods, corals and trilobites, and extinctions rarely occur in a rapid series of fast bursts.

Flora and fauna

The Silurian was the first period to see megafossils of extensive terrestrial biota, in the form of moss-like miniature forests along lakes and streams. However, the land fauna did not have a major impact on the Earth until it diversified in the Devonian.

The first fossil records of vascular plants, that is, land plants with tissues that carry water and food, appeared in the second half of the Silurian period. The earliest-known representatives of this group are *Cooksonia*. Most of the sediments containing *Cooksonia* are marine in nature. Preferred habitats were likely along rivers and streams. *Baragwanathia* appears to be almost as old, dating to the early Ludlow (420 million years) and has branching stems and needle-like leaves of 10–20 cm. The plant shows a high degree of development in relation to the age of its fossil remains. Fossils of this plant have been recorded in Australia, Canada and China. *Eohostimella heathana* is an early, probably terrestrial, "plant" known from compression fossils of early Silurian (Llandovery) age.[175] The chemistry of its fossils is similar to that of fossilised vascular plants, rather than algae.

The first bony fish, the Osteichthyes, appeared, represented by the Acanthodians covered with bony scales; fish reached considerable diversity and developed movable jaws, adapted from the supports of the front two or three gill arches. A diverse fauna of eurypterids (sea scorpions)—some of them several meters in length—prowled the shallow Silurian seas of North America; many of their fossils have been found in New York state. Leeches also made their appearance during the Silurian Period. Brachiopods, bryozoa, molluscs, hederelloids, tentaculitoids, crinoids and trilobites were abundant and diverse.Wikipedia:Citation needed Endobiotic symbionts were common in the corals and stromatoporoids.

Reef abundance was patchy; sometimes fossils are frequent but at other points are virtually absent from the rock record.

The earliest-known animals fully adapted to terrestrial conditions appear during the Mid-Silurian, including the millipede *Pneumodesmus*. Some evidence also suggests the presence of predatory trigonotarbid arachnoids and myriapods in Late Silurian facies. Predatory invertebrates would indicate that simple food webs were in place that included non-predatory prey animals. Extrapolating back from Early Devonian biota, Andrew Jeram *et al.* in 1990 suggested a food web based on as-yet-undiscovered detritivores and grazers on microorganisms.

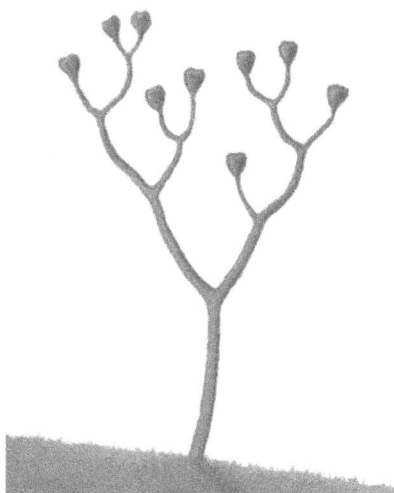

Figure 61: *Cooksonia, the earliest vascular plant, middle Silurian*

Figure 62: *Silurian sea bed fossils collected from Wren's Nest Nature Reserve, Dudley UK*

Figure 63: *Crinoid fragments in a Silurian (Pridoli) limestone (Saaremaa, Estonia)*

Figure 64: *Silurian sea bed fossils collected*
from Wren's Nest Nature Reserve, Dudley UK

Figure 65: *Artist's impression of Silurian underwater fauna*

Figure 66: *Eurypterus, a common Upper Silurian eurypterid*

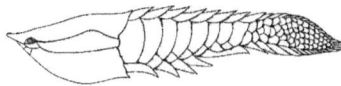

Figure 67: *Poraspis*

References

- Emiliani, Cesare. (1992). *Planet Earth : Cosmology, Geology, & the Evolution of Life & the Environment*. Cambridge University Press. (Paperback Edition ISBN 0-521-40949-7)
- Mikulic, DG, DEG Briggs, and J Kluessendorf. 1985. A new exceptionally preserved biota from the Lower Silurian of Wisconsin, USA. Philosophical Transactions of the Royal Society of London, 311B:75-86.
- Moore, RA, DEG Briggs, SJ Braddy, LI Anderson, DG Mikulic, and J Kluessendorf. 2005. A new synziphosurine (Chelicerata: Xiphosura) from the Late Llandovery (Silurian) Waukesha Lagerstatte, Wisconsin, USA. Journal of Paleontology:79(2), pp. 242–250.
- Ogg, Jim; June, 2004, *Overview of Global Boundary Stratotype Sections and Points (GSSP's)* https://web.archive.org/web/20060716071827/http://www.stratigraphy.org/gssp.htm Original version accessed April 30, 2006, redirected to archive on May 6, 2015.

External links

Wikisource has original works on the topic: *Paleozoic#Silurian*

Wikimedia Commons has media related to *Silurian*.

- Palaeos:[176] Silurian
- UCMP Berkeley:[177] The Silurian
- Paleoportal: Silurian strata in U.S., state by state[178]
- USGS:Silurian and Devonian Rocks (U.S.)[179]
- "International Commission on Stratigraphy (ICS)"[180]. *Geologic Time Scale 2004*. Retrieved September 19, 2005.
- Examples of Silurian Fossils[181]
- GeoWhen Database for the Silurian[182]

Devonian

Devonian Period *419.2–358.9 million years ago* PreЄ€ OSD C P T J K PgN	
Mean atmospheric O $_2$ content over period duration	c. 15 vol %[183,184] (75 % of modern level
Mean atmospheric CO $_2$ content over period duration	c. 2200 ppm[185] (8 times pre-industrial level)
Mean surface tempera- ture over period duration	c. 20 °C[186] (6 °C above modern level)
Sea level (above present day)	Relatively steady around 189m, gradu- ally falling to 120m through period

Events of the Devonian Period

Palæozoic
Key events of the Devonian Period.
Axis scale: millions of years ago.

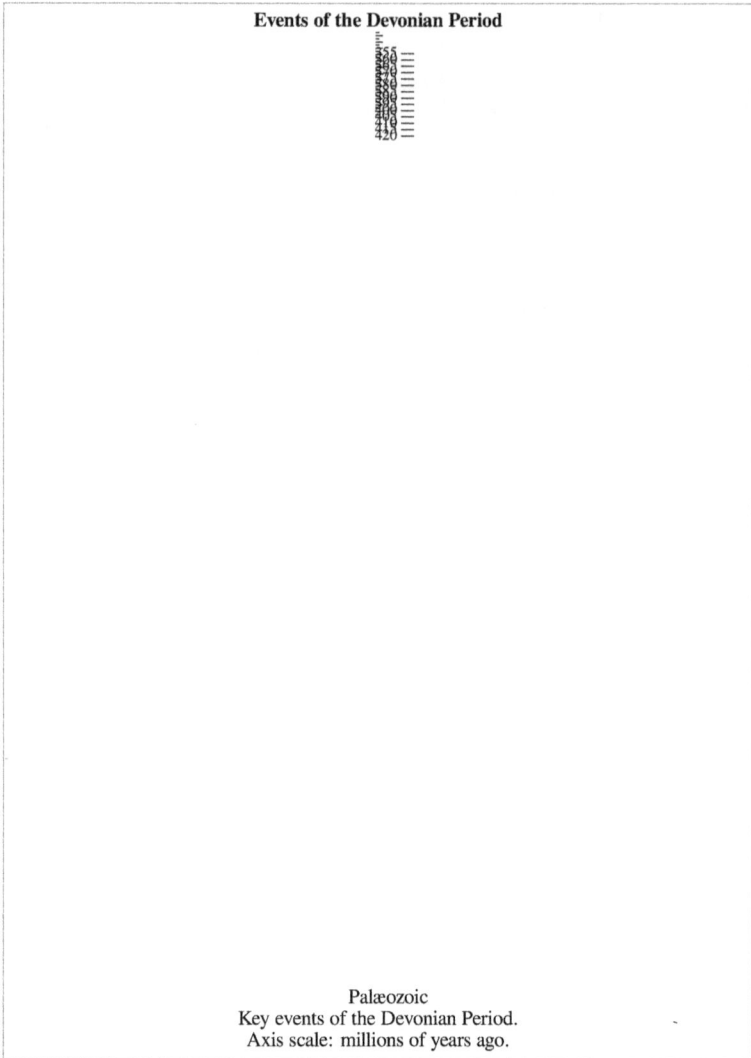

The **Devonian** is a geologic period and system of the Paleozoic, spanning 60 million years from the end of the Silurian, 419.2 million years ago (Mya), to the beginning of the Carboniferous, 358.9 Mya. It is named after Devon, England, where rocks from this period were first studied.

The first significant adaptive radiation of life on dry land occurred during the Devonian. Free-sporing vascular plants began to spread across dry land, forming extensive forests which covered the continents. By the middle of the Devonian, several groups of plants had evolved leaves and true roots, and by the end of the period the first seed-bearing plants appeared. Various terrestrial arthropods also became well-established.

Fish reached substantial diversity during this time, leading the Devonian to often be dubbed the "**Age of Fish**". The first ray-finned and lobe-finned bony fish appeared, while the placoderms began dominating almost every known aquatic environment. The ancestors of all four-limbed vertebrates (tetrapods) began adapting to walking on land, as their strong pectoral and pelvic fins gradually evolved into legs. In the oceans, primitive sharks became more numerous than in the Silurian and Late Ordovician.

The first ammonites, species of molluscs, appeared. Trilobites, the mollusc-like brachiopods and the great coral reefs, were still common. The Late Devonian extinction which started about 375 million years ago severely affected marine life, killing off all placodermi, and all trilobites, save for a few species of the order Proetida.

The palaeogeography was dominated by the supercontinent of Gondwana to the south, the continent of Siberia to the north, and the early formation of the small continent of Euramerica in between.

History

The period is named after Devon, a county in southwestern England, where a controversial argument in the 1830s over the age and structure of the rocks found distributed throughout the county was eventually resolved by the definition of the Devonian period in the geological timescale. The Great Devonian Controversy was a long period of vigorous argument and counter-argument between the main protagonists of Roderick Murchison with Adam Sedgwick against Henry De la Beche supported by George Bellas Greenough. Murchison and Sedgwick won the debate and named the period they proposed as the Devonian System.[187,188]

While the rock beds that define the start and end of the Devonian period are well identified, the exact dates are uncertain. According to the International Commission on Stratigraphy (Ogg, 2004), the Devonian extends from the end of the Silurian 419.2 Mya, to the beginning of the Carboniferous 358.9 Mya (in North America, the beginning of the Mississippian subperiod of the Carboniferous).

In nineteenth-century texts the Devonian has been called the "Old Red Age", after the red and brown terrestrial deposits known in the United Kingdom as the Old Red Sandstone in which early fossil discoveries were found. Another common term is "Age of the Fishes",[189] referring to the evolution of several major groups of fish that took place during the period. Older literature on the

Figure 68: *The rocks of Lummaton Quarry in Torquay in De-*
von played an early role in defining the Devonian period.

Anglo-Welsh basin divides it into the Downtonian, Dittonian, Breconian and
Farlovian stages, the latter three of which are placed in the Devonian.[190]

The Devonian has also erroneously been characterised as a "greenhouse age",
due to sampling bias: most of the early Devonian-age discoveries came from
the strata of western Europe and eastern North America, which at the time
straddled the Equator as part of the supercontinent of Euramerica where fossil
signatures of widespread reefs indicate tropical climates that were warm and
moderately humid but in fact the climate in the Devonian differed greatly during its epochs and between geographic regions. For example, during the Early
Devonian, arid conditions were prevalent through much of the world including
Siberia, Australia, North America, and China, but Africa and South America
had a warm temperate climate. In the Late Devonian, by contrast, arid conditions were less prevalent across the world and temperate climates were more
common.Wikipedia:Citation needed

Subdivisions

The Devonian Period is formally broken into Early, Middle and Late subdivisions. The rocks corresponding to those epochs are referred to as belonging to the Lower, Middle and Upper parts of the Devonian System.

Early Devonian

The Early Devonian lasted from $419.2^{191} \pm 2.8$ to $393.3^{192} \pm 2.5$ and began with the Lochkovian stage, which lasted until the Pragian. It spanned from $410.8^{193} \pm 2.8$ to $407.6^{194} \pm 2.5$, and was followed by the Emsian, which lasted until the Middle Devonian began, $393.3^{192} \pm 2.7$ million years ago.Wikipedia:Citation needed During this time, the first ammonoids appeared, descending from bactritoid nautiloids. Ammonoids during this time period were simple and differed little from their nautiloid counterparts. These ammonoids belong to the order Agoniatitida, which in later epochs evolved to new ammonoid orders, for example Goniatitida and Clymeniida. This class of cephalopod molluscs would dominate the marine fauna until the beginning of the Mesozoic era.

Middle Devonian

The Middle Devonian comprised two subdivisions: first the Eifelian, which then gave way to the Givetian $387.7^{195} \pm 2.7$ million years ago. During this time the jawless agnathan fishes began to decline in diversity in freshwater and marine environments partly due to drastic environmental changes and partly due to the increasing competition, predation and diversity of jawed fishes. The shallow, warm, oxygen-depleted waters of Devonian inland lakes, surrounded by primitive plants, provided the environment necessary for certain early fish to develop such essential characteristics as well developed lungs, and the ability to crawl out of the water and onto the land for short periods of time.Wikipedia:Citation needed

Late Devonian

Finally, the Late Devonian started with the Frasnian, $382.7^{196} \pm 2.8$ to $372.2^{197} \pm 2.5$, during which the first forests took shape on land. The first tetrapods appeared in the fossil record in the ensuing Famennian subdivision, the beginning and end of which are marked with extinction events. This lasted until the end of the Devonian, $358.9^{198} \pm 2.5$ million years ago.Wikipedia:Citation needed

Climate

The Devonian was a relatively warm period, and probably lacked any glaciers. The temperature gradient from the equator to the poles was not as large as it is today. The weather was also very arid, mostly along the equator where it was the driest. Reconstruction of tropical sea surface temperature from conodont apatite implies an average value of 30 °C (86 °F) in the Early Devonian. CO_2 levels dropped steeply throughout the Devonian period as the burial of the newly evolved forests drew carbon out of the atmosphere into sediments; this may be reflected by a Mid-Devonian cooling of around 5 °C (9 °F). The Late Devonian warmed to levels equivalent to the Early Devonian; while there is no corresponding increase in CO_2 concentrations, continental weathering increases (as predicted by warmer temperatures); further, a range of evidence, such as plant distribution, points to a Late Devonian warming.[199] The climate would have affected the dominant organisms in reefs; microbes would have been the main reef-forming organisms in warm periods, with corals and stromatoporoid sponges taking the dominant role in cooler times. The warming at the end of the Devonian may even have contributed to the extinction of the stromatoporoids.

Paleogeography

The Devonian period was a time of great tectonic activity, as Euramerica and Gondwana drew closer together.

The continent Euramerica (or Laurussia) was created in the early Devonian by the collision of Laurentia and Baltica, which rotated into the natural dry zone along the Tropic of Capricorn, which is formed as much in Paleozoic times as nowadays by the convergence of two great air-masses, the Hadley cell and the Ferrel cell. In these near-deserts, the Old Red Sandstone sedimentary beds formed, made red by the oxidised iron (hematite) characteristic of drought conditions.

Near the equator, the plate of Euramerica and Gondwana were starting to meet, beginning the early stages of the assembling of Pangaea. This activity further raised the northern Appalachian Mountains and formed the Caledonian Mountains in Great Britain and Scandinavia.

The west coast of Devonian North America, by contrast, was a passive margin with deep silty embayments, river deltas and estuaries, found today in Idaho and Nevada; an approaching volcanic island arc reached the steep slope of the continental shelf in Late Devonian times and began to uplift deep water deposits, a collision that was the prelude to the mountain-building episode at the beginning of the Carboniferous called the Antler orogeny.[200]

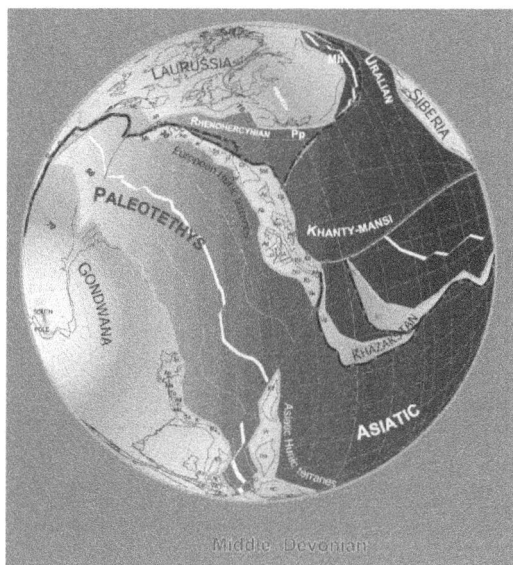

Figure 69: *The Paleo-Tethys Ocean opened during the Devonian*

Sea levels were high worldwide, and much of the land lay under shallow seas, where tropical reef organisms lived. The deep, enormous Panthalassa (the "universal ocean") covered the rest of the planet. Other minor oceans were the Paleo-Tethys Ocean, Proto-Tethys Ocean, Rheic Ocean, and Ural Ocean (which was closed during the collision with Siberia and Baltica).

Life

Marine biota

Sea levels in the Devonian were generally high. Marine faunas continued to be dominated by bryozoa, diverse and abundant brachiopods, the enigmatic hederellids, microconchids and corals. Lily-like crinoids (animals, their resemblance to flowers notwithstanding) were abundant, and trilobites were still fairly common. Among vertebrates, jawless armored fish (ostracoderms) declined in diversity, while the jawed fish (gnathostomes) simultaneously increased in both the sea and fresh water. Armored placoderms were numerous during the lower stages of the Devonian Period and became extinct in the Late Devonian, perhaps because of competition for food against the other fish species. Early cartilaginous (Chondrichthyes) and bony fishes (Osteichthyes) also become diverse and played a large role within the Devonian seas. The

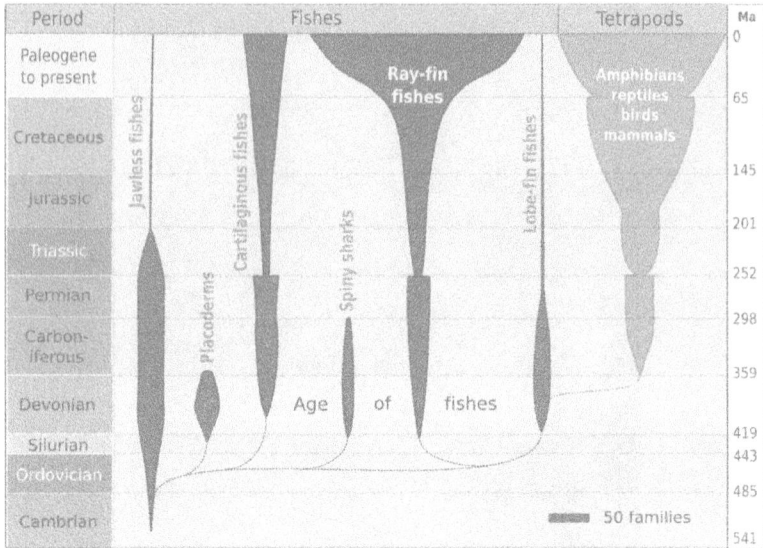

Figure 70: *Spindle diagram for the evolution of fish and other vertebrate classes. The diagram is based on Michael Benton, 2005.*[201]

first abundant genus of shark, *Cladoselache*, appeared in the oceans during the Devonian Period. The great diversity of fish around at the time has led to the Devonian being given the name "The Age of Fish" in popular culture.

The first ammonites also appeared during or slightly before the early Devonian Period around 400 Mya.[202]

Reefs

A now dry barrier reef, located in present-day Kimberley Basin of northwest Australia, once extended a thousand kilometres, fringing a Devonian continent. Reefs in general are built by various carbonate-secreting organisms that have the ability to erect wave-resistant frameworks close to sea level. The main contributors of the Devonian reefs were unlike modern reefs, which are constructed mainly by corals and calcareous algae. They were composed of calcareous algae and coral-like stromatoporoids, and tabulate and rugose corals, in that order of importance.Wikipedia:Please clarify

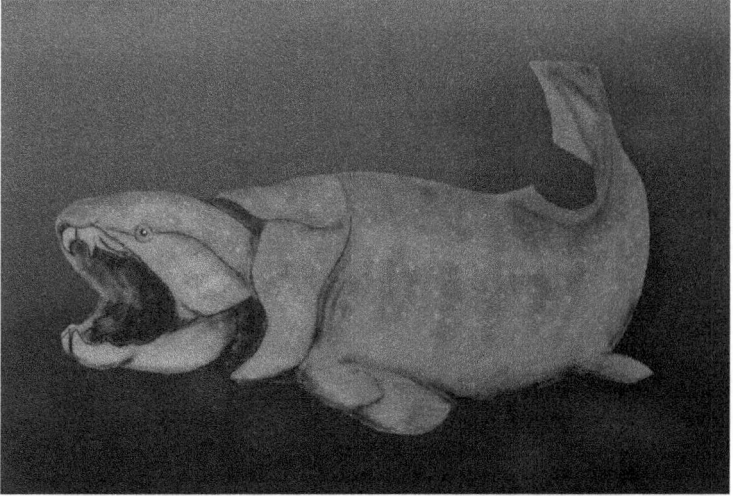

Figure 71: *Dunkleosteus, one of the largest armoured fish ever to roam the planet, lived during the late Devonian*

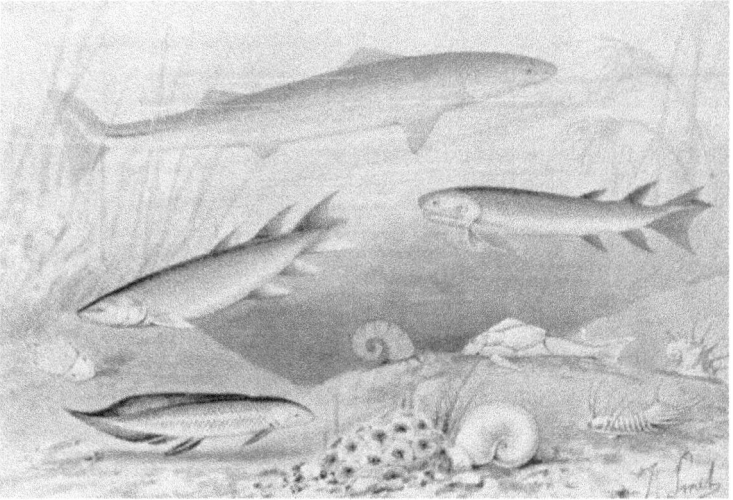

Figure 72: *Early shark Cladoselache, several lobe-finned fishes, including Eusthenopteron that was an early marine tetrapod, and the placoderm Bothriolepis in a painting from 1905*

Figure 73: *Enrolled phacopid trilobite from the Devonian of Ohio*

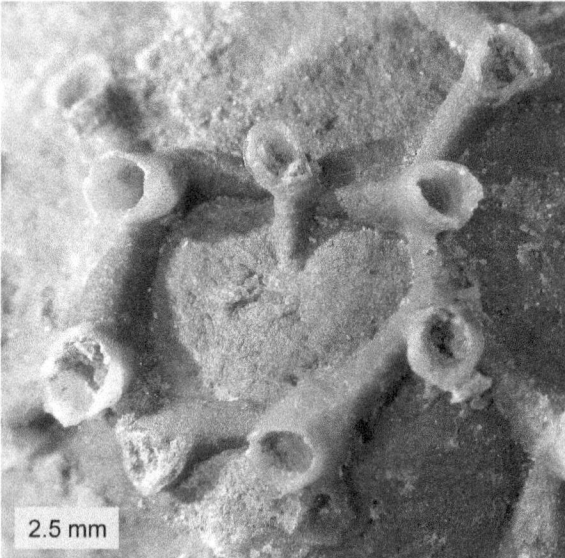

Figure 74: *The common tabulate coral Aulopora from the Middle Devonian of Ohio – view of colony encrusting a brachiopod valve*

Figure 75: *Tropidoleptus carinatus, an orthid bra-chiopod from the Middle Devonian of New York.*

Figure 76: *Pleurodictyum americanum,
Kashong Shale, Middle Devonian of New York*

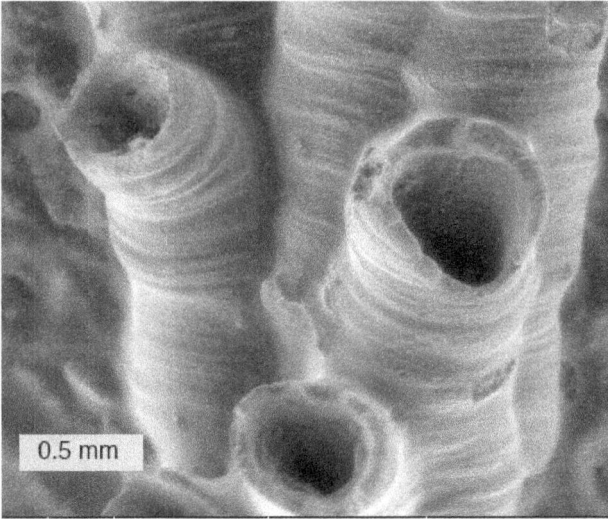

Figure 77: *SEM image of a hederelloid from the Devonian of Michigan (largest tube diameter is 0.75 mm)*

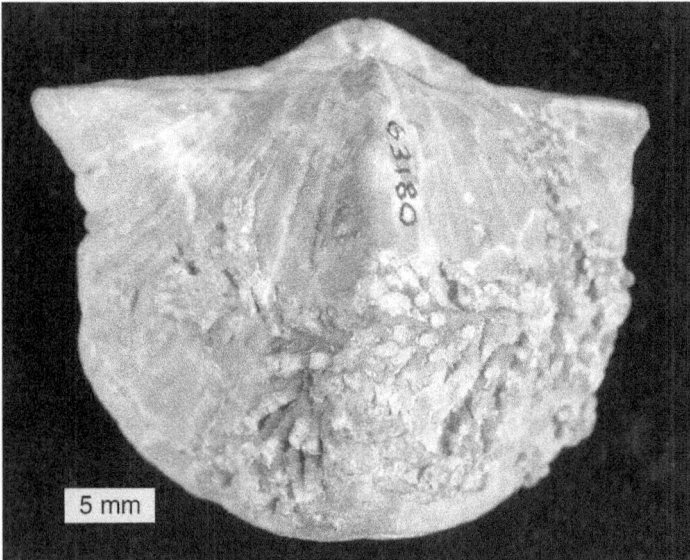

Figure 78: *Devonian spiriferid brachiopod from Ohio which served as a host substrate for a colony of hederelloids*

Figure 79: *The Devonian period marks the beginning of extensive land colonisation by plants. With large land-dwelling herbivores not yet present, large forests grew and shaped the landscape.*

Terrestrial biota

By the Devonian Period, life was well underway in its colonisation of the land. The moss forests and bacterial and algal mats of the Silurian were joined early in the period by primitive rooted plants that created the first stable soils and harbored arthropods like mites, scorpions, trigonotarbids and myriapods (although arthropods appeared on land much earlier than in the Early Devonian and the existence of fossils such as *Climactichnites* suggest that land arthropods may have appeared as early as the Cambrian). Also the first possible fossils of insects appeared around 416 Mya in the Early Devonian. Evidence for the earliest tetrapods takes the form of trace fossils in shallow lagoon environments within a marine carbonate platform/shelf during the Middle Devonian, although these traces have been questioned and an interpretation as fish feeding traces (Piscichnus) has been advanced.

The greening of land

Many Early Devonian plants did not have true roots or leaves like extant plants although vascular tissue is observed in many of those plants. Some of the early land plants such as *Drepanophycus* likely spread by vegetative growth and spores. The earliest land plants such as *Cooksonia* consisted of leafless, dichotomous axes and terminal sporangia and were generally very short-statured, and grew hardly more than a few centimetres tall. By far the largest land organism during this period was the enigmatic *Prototaxites*, which was possibly the fruiting body of an enormous fungus, rolled liverwort mat, or another organism of uncertain affinities that stood more than 8 metres tall, and towered over the low, carpet-like vegetation. By the Middle Devonian, shrub-like forests of primitive plants existed: lycophytes, horsetails, ferns, and progymnosperms had evolved. Most of these plants had true roots and leaves, and many were quite tall. The earliest-known trees, from the genus *Wattieza*, appeared in the Late Devonian around 385 Mya. In the Late Devonian, the tree-like ancestral Progymnosperm *Archaeopteris* which had conifer-like true wood and fern-like foliage and the cladoxylopsids grew.[203] (See also: lignin.) These are the oldest-known trees of the world's first forests. By the end of the Devonian, the first seed-forming plants had appeared. This rapid appearance of so many plant groups and growth forms has been called the "Devonian Explosion".

The 'greening' of the continents acted as a carbon sink, and atmospheric concentrations of carbon dioxide may have dropped. This may have cooled the climate and led to a massive extinction event. See Late Devonian extinction.

Animals and the first soils

Primitive arthropods co-evolved with this diversified terrestrial vegetation structure. The evolving co-dependence of insects and seed-plants that characterised a recognisably modern world had its genesis in the Late Devonian period. The development of soils and plant root systems probably led to changes in the speed and pattern of erosion and sediment deposition. The rapid evolution of a terrestrial ecosystem that contained copious animals opened the way for the first vertebrates to seek out a terrestrial living. By the end of the Devonian, arthropods were solidly established on the land.

Late Devonian extinction

A major extinction occurred at the beginning of the last phase of the Devonian period, the Famennian faunal stage (the Frasnian-Famennian boundary), about 372.2 Mya, when all the fossil agnathan fishes, save for the psammosteid heterostraci, suddenly disappeared. A second strong pulse closed the Devonian period. The Late Devonian extinction was one of five major extinction events in the history of the Earth's biota, and was more drastic than the familiar extinction event that closed the Cretaceous.

Marine Genus Biodiversity: Extinction Intensity

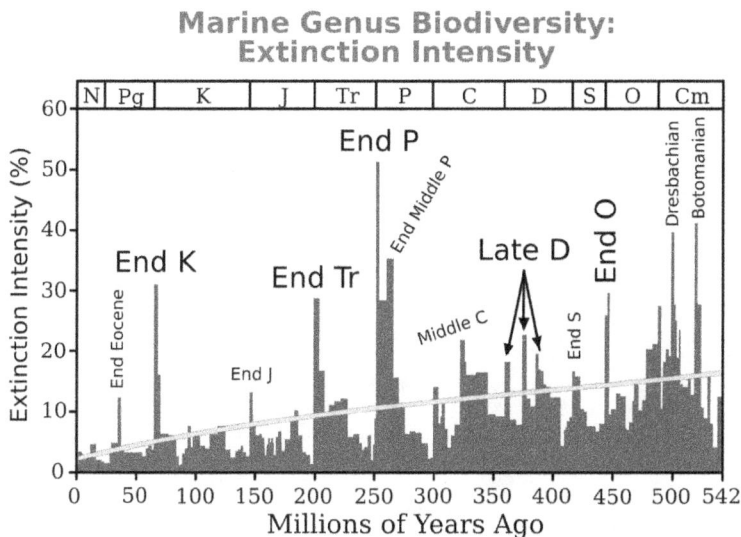

Figure 80: *The Late Devonian is charac-terised by three episodes of extinction ("Late D")*

The Devonian extinction crisis primarily affected the marine community, and selectively affected shallow warm-water organisms rather than cool-water organisms. The most important group to be affected by this extinction event were the reef-builders of the great Devonian reef systems.

Amongst the severely affected marine groups were the brachiopods, trilobites, ammonites, conodonts, and acritarchs, as well as jawless fish, and all placoderms. Land plants as well as freshwater species, such as our tetrapod ancestors, were relatively unaffected by the Late Devonian extinction event (there is a counterargument that the Devonian extinctions nearly wiped out the tetrapods[204]).

The reasons for the Late Devonian extinctions are still unknown, and all explanations remain speculative.[205] Canadian paleontologist Digby McLaren suggested in 1969 that the Devonian extinction events were caused by an asteroid impact. However, while there were Late Devonian collision events (see the Alamo bolide impact), little evidence supports the existence of a large enough Devonian crater.

References

General

- Ogg, Jim; June, 2004, *Overview of Global Boundary Stratotype Sections and Points (GSSP's)* https://web.archive.org/web/20060716071827/http://www.stratigraphy.org/gssp.htm Accessed April 30, 2006.
- Palaeos[206]
- Age of Fishes Museum[207]

External links

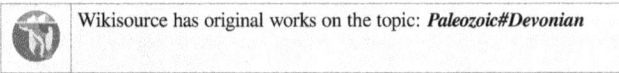

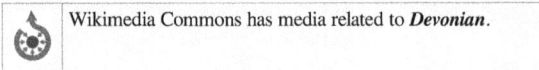

Wikisource has original works on the topic: *Paleozoic#Devonian*

Wikimedia Commons has media related to *Devonian*.

- The Devonian times[208] - an excellent and frequently updated resource focussing on the Devonian period
- UC Berkeley site introduces the Devonian.[209]
- "International Commission on Stratigraphy (ICS)"[210]. *Geologic Time Scale 2004*. Retrieved September 19, 2005.
- Examples of Devonian Fossils[211]

Carboniferous

Carboniferous Period *358.9–298.9 million years ago* Pre€€ OSD **C** P T J K P₃N	
Mean atmospheric O₂ content over period duration	c. 32.3 vol %[212,213] (162 % of modern level
Mean atmospheric CO₂ content over period duration	c. 800 ppm[214] (3 times pre-industrial level)
Mean surface temperature over period duration	c. 14 °C[215] (0 °C above modern level)
Sea level (above present day)	Falling from 120 m to present-day level throughout the Mississippian, then rising steadily to about 80 m at end of period

Key events in the Carboniferous

Key events of the Carboniferous Period
Axis scale: millions of years ago

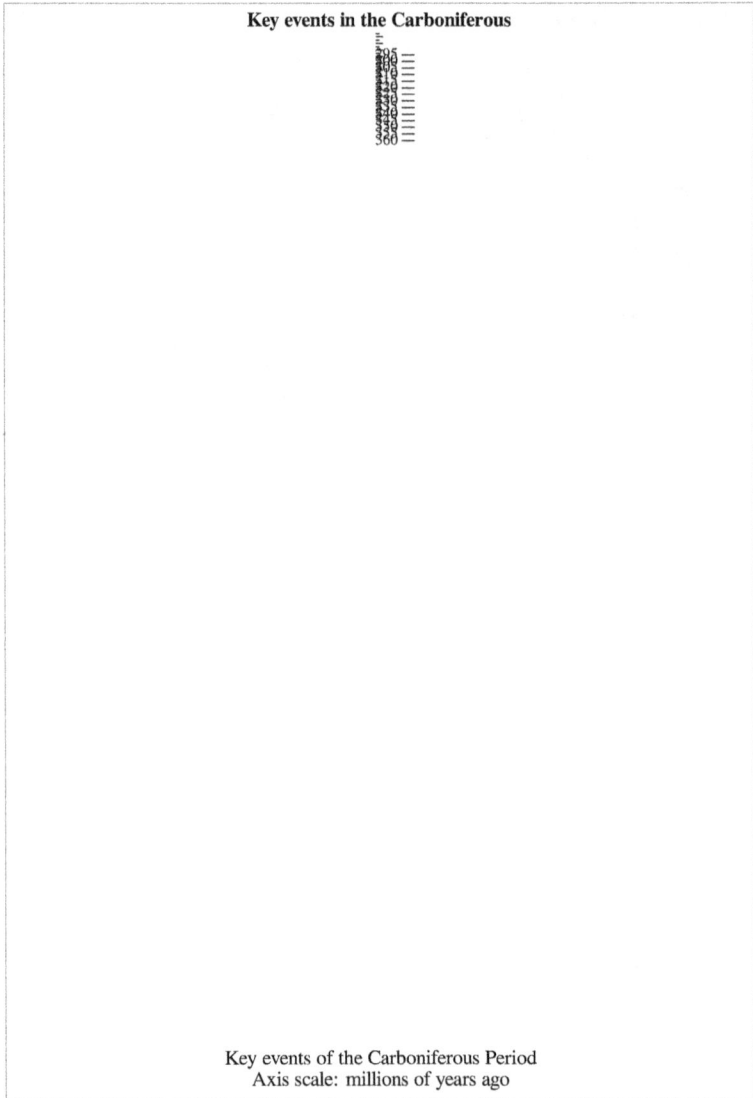

The **Carboniferous** is a geologic period and system that spans 60 million years from the end of the Devonian Period 358.9 million years ago (Mya), to the beginning of the Permian Period, 298.9 Mya. The name *Carboniferous* means "coal-bearing" and derives from the Latin words *carbō* ("coal") and *ferō* ("I bear, I carry"), and was coined by geologists William Conybeare and William Phillips in 1822.[216]

Based on a study of the British rock succession, it was the first of the modern 'system' names to be employed, and reflects the fact that many coal beds were formed globally during that time.[217] The Carboniferous is often treated in North America as two geological periods, the earlier Mississippian and the later Pennsylvanian. Terrestrial animal life was well established by the Carboniferous period. Amphibians were the dominant land vertebrates, of which one branch would eventually evolve into amniotes, the first solely terrestrial vertebrates.

Arthropods were also very common, and many (such as *Meganeura*) were much larger than those of today. Vast swaths of forest covered the land, which would eventually be laid down and become the coal beds characteristic of the Carboniferous stratigraphy evident today. The atmospheric content of oxygen also reached its highest levels in geological history during the period, 35% compared with 21% today, allowing terrestrial invertebrates to evolve to great size.

The later half of the period experienced glaciations, low sea level, and mountain building as the continents collided to form Pangaea. A major marine and terrestrial extinction event, the Carboniferous rainforest collapse, occurred at the end of the period, caused by climate change.

Subdivisions

In the United States the Carboniferous is usually broken into Mississippian (earlier) and Pennsylvanian (later) subperiods. The Mississippian is about twice as long as the Pennsylvanian, but due to the large thickness of coal-bearing deposits with Pennsylvanian ages in Europe and North America, the two subperiods were long thought to have been more or less equal in duration.[218]

System	Series (NW Europe)	Stage (NW Europe)	Series (ICS)	Stage (ICS)	Age (Ma)
Permian					younger
Carboniferous	Silesian	Stephanian	Pennsylvanian	Gzhelian	298.9–303.7
		Westphalian		Kasimovian	303.7–307.0
				Moscovian	307.0–315.2
				Bashkirian	315.2–323.2
		Namurian			
			Mississippian	Serpukhovian	323.2–330.9

Dinantian	Visean		Visean	330.9–346.7
	Tournaisian		Tournaisian	346.7–358.9
Devonian				older

Subdivisions of the Carboniferous system in Europe compared with the official ICS-stages (as of 2018)

In Europe the Lower Carboniferous sub-system is known as the Dinantian, comprising the Tournaisian and Visean Series, dated at 362.5-332.9 Ma, and the Upper Carboniferous sub-system is known as the Silesian, comprising the Namurian, Westphalian, and Stephanian Series, dated at 332.9-298.9 Ma. The Silesian is roughly contemporaneous with the late Mississippian Serpukhovian plus the Pennsylvanian. In Britain the Dinantian is traditionally known as the Carboniferous Limestone, the Namurian as the Millstone Grit, and the West-phalian as the Coal Measures and Pennant Sandstone.

The International Commission on Stratigraphy (ICS) faunal stages (in bold) from youngest to oldest, together with some of their regional subdivisions, are:

Late Pennsylvanian: Gzhelian (most recent)

- Noginskian / Virgilian *(part)*

Late Pennsylvanian: Kasimovian

- Klazminskian
- Dorogomilovksian / Virgilian *(part)*
- Chamovnicheskian / Cantabrian / Missourian
- Krevyakinskian / Cantabrian / Missourian

Middle Pennsylvanian: Moscovian

- Myachkovskian / Bolsovian / Desmoinesian
- Podolskian / Desmoinesian
- Kashirskian / Atokan
- Vereiskian / Bolsovian / Atokan

Early Pennsylvanian: Bashkirian / Morrowan

- Melekesskian / Duckmantian
- Cheremshanskian / Langsettian
- Yeadonian
- Marsdenian
- Kinderscoutian

Late Mississippian: Serpukhovian

- Alportian
- Chokierian / Chesterian / Elvirian

- Arnsbergian / Elvirian
- Pendleian

Middle Mississippian: Visean

- Brigantian / St Genevieve / Gasperian / Chesterian
- Asbian / Meramecian
- Holkerian / Salem
- Arundian / Warsaw / Meramecian
- Chadian / Keokuk / Osagean *(part)* / Osage *(part)*

Early Mississippian: Tournaisian (oldest)

- Ivorian / *(part)* / Osage *(part)*
- Hastarian / Kinderhookian / Chouteau

Palaeogeography

A global drop in sea level at the end of the Devonian reversed early in the Carboniferous; this created the widespread inland seas and the carbonate deposition of the Mississippian. There was also a drop in south polar temperatures; southern Gondwanaland was glaciated throughout the period, though it is uncertain if the ice sheets were a holdover from the Devonian or not. These conditions apparently had little effect in the deep tropics, where lush swamps, later to become coal, flourished to within 30 degrees of the northernmost glaciers.

Mid-Carboniferous, a drop in sea level precipitated a major marine extinction, one that hit crinoids and ammonites especially hard. This sea level drop and the associated unconformity in North America separate the Mississippian subperiod from the Pennsylvanian subperiod. This happened about 323 million years ago, at the onset of the Permo-Carboniferous Glaciation.Wikipedia:Citation needed

The Carboniferous was a time of active mountain-building, as the supercontinent Pangaea came together. The southern continents remained tied together in the supercontinent Gondwana, which collided with North America–Europe (Laurussia) along the present line of eastern North America. This continental collision resulted in the Hercynian orogeny in Europe, and the Alleghenian orogeny in North America; it also extended the newly uplifted Appalachians southwestward as the Ouachita Mountains. In the same time frame, much of present eastern Eurasian plate welded itself to Europe along the line of the Ural Mountains. Most of the Mesozoic supercontinent of Pangea was now assembled, although North China (which would collide in the Latest Carboniferous), and South China continents were still separated from Laurasia. The Late Carboniferous Pangaea was shaped like an "O."

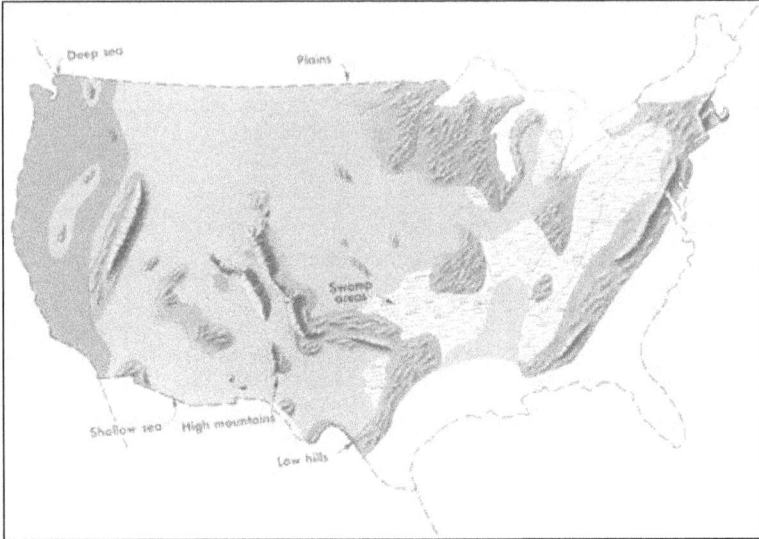

Figure 81: *Generalized geographic map of the*
United States in Middle Pennsylvanian time.

There were two major oceans in the Carboniferous—Panthalassa and Paleo-Tethys, which was inside the "O" in the Carboniferous Pangaea. Other minor oceans were shrinking and eventually closed - Rheic Ocean (closed by the assembly of South and North America), the small, shallow Ural Ocean (which was closed by the collision of Baltica and Siberia continents, creating the Ural Mountains) and Proto-Tethys Ocean (closed by North China collision with Siberia/Kazakhstania).

Climate

Average global temperatures in the Early Carboniferous Period were high: approximately 20 °C (68 °F). However, cooling during the Middle Carboniferous reduced average global temperatures to about 12 °C (54 °F). Lack of growth rings of fossilized trees suggest a lack of seasons of a tropical climate. Glaciations in Gondwana, triggered by Gondwana's southward movement, continued into the Permian and because of the lack of clear markers and breaks, the deposits of this glacial period are often referred to as Permo-Carboniferous in age.

The thicker atmosphere and stronger coriolis effect due to Earth's faster rotation (a day lasted for 22.4 hours in early Carboniferous) created significantly stronger winds than today.

Figure 82: *Lower Carboniferous marble in Big Cottonwood Canyon, Wasatch Mountains, Utah.*

The cooling and drying of the climate led to the Carboniferous Rainforest Collapse (CRC) during the late Carboniferous. Tropical rainforests fragmented and then were eventually devastated by climate change.

Rocks and coal

Carboniferous rocks in Europe and eastern North America largely consist of a repeated sequence of limestone, sandstone, shale and coal beds.[219] In North America, the early Carboniferous is largely marine limestone, which accounts for the division of the Carboniferous into two periods in North American schemes. The Carboniferous coal beds provided much of the fuel for power generation during the Industrial Revolution and are still of great economic importance.

The large coal deposits of the Carboniferous may owe their existence primarily to two factors. The first of these is the appearance of wood tissue and bark-bearing trees. The evolution of the wood fiber lignin and the bark-sealing, waxy substance suberin variously opposed decay organisms so effectively that dead materials accumulated long enough to fossilise on a large scale. The second factor was the lower sea levels that occurred during the Carboniferous as compared to the preceding Devonian period. This promoted the development of extensive lowland swamps and forests in North America and Europe. Based on a genetic analysis of mushroom fungi, it was proposed that large quantities of wood were buried during this period because animals and decomposing bacteria had not yet evolved enzymes that could effectively digest the resistant phenolic lignin polymers and waxy suberin polymers. They suggest that fungi that could break those substances down effectively only became dominant towards the end of the period, making subsequent coal formation much rarer.

The Carboniferous trees made extensive use of lignin. They had bark to wood ratios of 8 to 1, and even as high as 20 to 1. This compares to modern values less than 1 to 4. This bark, which must have been used as support as well as protection, probably had 38% to 58% lignin. Lignin is insoluble, too large to pass through cell walls, too heterogeneous for specific enzymes, and toxic, so that few organisms other than Basidiomycetes fungi can degrade it. To oxidize it requires an atmosphere of greater than 5% oxygen, or compounds such as peroxides. It can linger in soil for thousands of years and its toxic breakdown products inhibit decay of other substances.[220] One possible reason for its high percentages in plants at that time was to provide protection from insects in a world containing very effective insect herbivores (but nothing remotely as effective as modern insectivores) and probably many fewer protective toxins produced naturally by plants than exist today. As a result, undegraded carbon built up, resulting in the extensive burial of biologically fixed carbon, leading to an increase in oxygen levels in the atmosphere; estimates place the peak oxygen content as high as 35%, as compared to 21% today. This oxygen level may have increased wildfire activity. It also may have promoted gigantism of insects and amphibians — creatures that have been constrained in size by respiratory systems that are limited in their physiological ability to transport and distribute oxygen at the lower atmospheric concentrations that have since been available.

In eastern North America, marine beds are more common in the older part of the period than the later part and are almost entirely absent by the late Carboniferous. More diverse geology existed elsewhere, of course. Marine life is especially rich in crinoids and other echinoderms. Brachiopods were abundant. Trilobites became quite uncommon. On land, large and diverse plant populations existed. Land vertebrates included large amphibians.

Figure 83: *Etching depicting some of the most significant plants of the Carboniferous.*

Life

Plants

Wikisourcehas the text of the 1879 *American Cyclopædia*article **Coal Plants**.

Figure 84: *Ancient in situ lycopsid, probably Sigillaria, with attached stigmarian roots.*

Early Carboniferous land plants, some of which were preserved in coal balls, were very similar to those of the preceding Late Devonian, but new groups also appeared at this time.

The main Early Carboniferous plants were the Equisetales (horse-tails), Sphenophyllales (scrambling plants), Lycopodiales (club mosses), Lepido-dendrales (scale trees), Filicales (ferns), Medullosales (informally included in the "seed ferns", an artificial assemblage of a number of early gymnosperm groups) and the Cordaitales. These continued to dominate throughout the pe-riod, but during late Carboniferous, several other groups, Cycadophyta (cy-cads), the Callistophytales (another group of "seed ferns"), and the Voltziales (related to and sometimes included under the conifers), appeared.

The Carboniferous lycophytes of the order Lepidodendrales, which are cousins (but not ancestors) of the tiny club-moss of today, were huge trees with trunks 30 meters high and up to 1.5 meters in diameter. These included *Lepido-dendron* (with its cone called Lepidostrobus), *Anabathra*, *Lepidophloios* and *Sigillaria*. The roots of several of these forms are known as Stigmaria. Unlike present-day trees, their secondary growth took place in the cortex, which also provided stability, instead of the xylem.[221] The Cladoxylopsids were large trees, that were ancestors of ferns, first arising in the Carboniferous.[222]

Figure 85: *Base of a lycopsid showing connection with bifurcating stigmarian roots.*

The fronds of some Carboniferous ferns are almost identical with those of living species. Probably many species were epiphytic. Fossil ferns and "seed ferns" include *Pecopteris, Cyclopteris, Neuropteris, Alethopteris,* and *Sphenopteris; Megaphyton* and *Caulopteris* were tree ferns.

The Equisetales included the common giant form *Calamites*, with a trunk diameter of 30 to 60 cm (24 in) and a height of up to 20 m (66 ft). *Sphenophyllum* was a slender climbing plant with whorls of leaves, which was probably related both to the calamites and the lycopods.

Cordaites, a tall plant (6 to over 30 meters) with strap-like leaves, was related to the cycads and conifers; the catkin-like reproductive organs, which bore ovules/seeds, is called *Cardiocarpus*. These plants were thought to live in swamps. True coniferous trees (*Walchia*, of the order Voltziales) appear later in the Carboniferous, and preferred higher drier ground.

Marine invertebrates

In the oceans the most important marine invertebrate groups are the Foraminifera, corals, Bryozoa, Ostracoda, brachiopods, ammonoids, hederelloids, microconchids and echinoderms (especially crinoids). For the first time foraminifera take a prominent part in the marine faunas. The large spindle-shaped genus *Fusulina* and its relatives were abundant in what is now Russia,

China, Japan, North America; other important genera include *Valvulina*, *Endothyra*, *Archaediscus*, and *Saccammina* (the latter common in Britain and Belgium). Some Carboniferous genera are still extant.

The microscopic shells of radiolarians are found in cherts of this age in the Culm of Devon and Cornwall, and in Russia, Germany and elsewhere. Sponges are known from spicules and anchor ropes, and include various forms such as the Calcispongea *Cotyliscus* and *Girtycoelia*, the demosponge *Chaetetes*, and the genus of unusual colonial glass sponges *Titusvillia*.

Both reef-building and solitary corals diversify and flourish; these include both rugose (for example, *Caninia*, *Corwenia*, *Neozaphrentis*), heterocorals, and tabulate (for example, *Chladochonus*, *Michelinia*) forms. Conularids were well represented by *Conularia*

Bryozoa are abundant in some regions; the fenestellids including *Fenestella*, *Polypora*, and *Archimedes*, so named because it is in the shape of an Archimedean screw. Brachiopods are also abundant; they include productids, some of which (for example, *Gigantoproductus*) reached very large (for brachiopods) size and had very thick shells, while others like *Chonetes* were more conservative in form. Athyridids, spiriferids, rhynchonellids, and terebratulids are also very common. Inarticulate forms include *Discina* and *Crania*. Some species and genera had a very wide distribution with only minor variations.

Annelids such as *Serpulites* are common fossils in some horizons. Among the mollusca, the bivalves continue to increase in numbers and importance. Typical genera include *Aviculopecten*, *Posidonomya*, *Nucula*, *Carbonicola*, *Edmondia*, and *Modiola* Gastropods are also numerous, including the genera *Murchisonia*, *Euomphalus*, *Naticopsis*. Nautiloid cephalopods are represented by tightly coiled nautilids, with straight-shelled and curved-shelled forms becoming increasingly rare. Goniatite ammonoids are common.

Trilobites are rarer than in previous periods, on a steady trend towards extinction, represented only by the proetid group. Ostracoda, a class of crustaceans, were abundant as representatives of the meiobenthos; genera included *Amphissites*, *Bairdia*, *Beyrichiopsis*, *Cavellina*, *Coryellina*, *Cribroconcha*, *Hollinella*, *Kirkbya*, *Knoxiella*, and *Libumella*.

Amongst the echinoderms, the crinoids were the most numerous. Dense submarine thickets of long-stemmed crinoids appear to have flourished in shallow seas, and their remains were consolidated into thick beds of rock. Prominent genera include *Cyathocrinus*, *Woodocrinus*, and *Actinocrinus*. Echinoids such as *Archaeocidaris* and *Palaeechinus* were also present. The blastoids, which included the Pentreinitidae and Codasteridae and superficially resembled crinoids in the possession of long stalks attached to the seabed, attain their maximum development at this time.

Figure 86: *Aviculopecten subcardiformis; a bivalve from the Logan Formation (Lower Carboniferous) of Wooster, Ohio (external mold).*

Figure 87: *Bivalves (Aviculopecten) and brachiopods (Syringothyris) in the Logan Formation (Lower Carboniferous) in Wooster, Ohio.*

Figure 88: *Syringothyris sp.; a spiriferid brachiopod from the Logan Formation (Lower Carboniferous) of Wooster, Ohio (internal mold).*

Figure 89: *Palaeophycus ichnosp.; a trace fossil from the Logan Formation (Lower Carboniferous) of Wooster, Ohio.*

Figure 90: *Crinoid calyx from the Lower Carboniferous of Ohio with a conical platyceratid gastropod (Palaeocapulus acutirostre) attached.*

Figure 91: *Conulariid from the Lower Carboniferous of Indiana.*

Figure 92: *Tabulate coral (a syringoporid); Boone Lime-stone (Lower Carboniferous) near Hiwasse, Arkansas.*

Freshwater and lagoonal invertebrates

Freshwater Carboniferous invertebrates include various bivalve molluscs that lived in brackish or fresh water, such as *Anthraconaia, Naiadites*, and *Carbonicola*; diverse crustaceans such as *Candona, Carbonita, Darwinula, Estheria, Acanthocaris, Dithyrocaris*, and *Anthrapalaemon*.

The Eurypterids were also diverse, and are represented by such genera as *Anthraconectes, Megarachne* (originally misinterpreted as a giant spider, hence its name) and the specialised very large *Hibbertopterus*. Many of these were amphibious.

Frequently a temporary return of marine conditions resulted in marine or brackish water genera such as *Lingula*, Orbiculoidea, and *Productus* being found in the thin beds known as marine bands.

Terrestrial invertebrates

Fossil remains of air-breathing insects, myriapods and arachnids are known from the late Carboniferous, but so far not from the early Carboniferous. The first true priapulids appeared during this period. Their diversity when they do appear, however, shows that these arthropods were both well developed and numerous. Their large size can be attributed to the moistness of the environment (mostly swampy fern forests) and the fact that the oxygen concentration in the Earth's atmosphere in the Carboniferous was much higher than today. This required less effort for respiration and allowed arthropods to grow larger with the up to 2.6-meter-long (8.5 ft) millipede-like *Arthropleura* being the

Figure 93: *The upper Carboniferous giant spider-like eu-rypterid Megarachne grew to legspans of 50 cm (20 in).*

largest-known land invertebrate of all time. Among the insect groups are the huge predatory Protodonata (griffinflies), among which was *Meganeura*, a giant dragonfly-like insect and with a wingspan of ca. 75 cm (30 in)—the largest flying insect ever to roam the planet. Further groups are the Syntonopterodea (relatives of present-day mayflies), the abundant and often large sap-sucking Palaeodictyopteroidea, the diverse herbivorous Protorthoptera, and numerous basal Dictyoptera (ancestors of cockroaches). Many insects have been obtained from the coalfields of Saarbrücken and Commentry, and from the hollow trunks of fossil trees in Nova Scotia. Some British coalfields have yielded good specimens: *Archaeoptitus*, from the Derbyshire coalfield, had a spread of wing extending to more than 35 cm (14 in); some specimens (*Brodia*) still exhibit traces of brilliant wing colors. In the Nova Scotian tree trunks land snails (*Archaeozonites, Dendropupa*) have been found.

Figure 94: *The late Carboniferous giant dragonfly-like insect Meganeura grew to wingspans of .*

Figure 95: *The gigantic Pulmonoscorpius from the early Carboniferous reached a length of up to .*

Fish

Many fish inhabited the Carboniferous seas; predominantly Elasmobranchs (sharks and their relatives). These included some, like *Psammodus*, with crushing pavement-like teeth adapted for grinding the shells of brachiopods, crustaceans, and other marine organisms. Other sharks had piercing teeth, such as the Symmoriida; some, the petalodonts, had peculiar cycloid cutting teeth. Most of the sharks were marine, but the Xenacanthida invaded fresh waters of the coal swamps. Among the bony fish, the Palaeonisciformes found in coastal waters also appear to have migrated to rivers. Sarcopterygian fish were also prominent, and one group, the Rhizodonts, reached very large size.

Most species of Carboniferous marine fish have been described largely from teeth, fin spines and dermal ossicles, with smaller freshwater fish preserved whole.

Freshwater fish were abundant, and include the genera *Ctenodus*, *Uronemus*, *Acanthodes*, *Cheirodus*, and *Gyracanthus*.

Sharks (especially the *Stethacanthids*) underwent a major evolutionary radiation during the Carboniferous. It is believed that this evolutionary radiation occurred because the decline of the placoderms at the end of the Devonian period caused many environmental niches to become unoccupied and allowed new organisms to evolve and fill these niches. As a result of the evolutionary radiation Carboniferous sharks assumed a wide variety of bizarre shapes including *Stethacanthus* which possessed a flat brush-like dorsal fin with a patch of denticles on its top. *Stethacanthus*'s unusual fin may have been used in mating rituals.

Figure 96: *Akmonistion of the shark order Symmori-ida roamed the oceans of the early Carboniferous.*

Figure 97: *Falcatus was a Carboniferous shark, with a high degree of sexual dimorphism.*

Tetrapods

Carboniferous amphibians were diverse and common by the middle of the period, more so than they are today; some were as long as 6 meters, and those fully terrestrial as adults had scaly skin.[223] They included a number of basal tetrapod groups classified in early books under the Labyrinthodontia. These had long bodies, a head covered with bony plates and generally weak or undeveloped limbs. The largest were over 2 meters long. They were accompanied by an assemblage of smaller amphibians included under the Lepospondyli, often only about 15 cm (6 in) long. Some Carboniferous amphibians were aquatic and lived in rivers (*Loxomma, Eogyrinus, Proterogyrinus*); others may have been semi-aquatic (*Ophiderpeton, Amphibamus, Hyloplesion*) or terrestrial (*Dendrerpeton, Tuditanus, Anthracosaurus*).

The Carboniferous Rainforest Collapse slowed the evolution of amphibians who could not survive as well in the cooler, drier conditions. Reptiles, however, prospered due to specific key adaptations. One of the greatest evolutionary innovations of the Carboniferous was the amniote egg, which allowed the laying of eggs in a dry environment, allowing for the further exploitation of the land by certain tetrapods. These included the earliest sauropsid reptiles (*Hylonomus*), and the earliest known synapsid (*Archaeothyris*). These small lizard-like animals quickly gave rise to many descendants, reptiles, birds, and mammals.

Reptiles underwent a major evolutionary radiation in response to the drier climate that preceded the rainforest collapse.[224] By the end of the Carboniferous period, amniotes had already diversified into a number of groups, including protorothyridids, captorhinids, araeoscelids, and several families of pelycosaurs.

Figure 98: *The amphibian-like Pederpes, the most primitive Mississippian tetrapod*

Figure 99: *Hylonomus, the earliest sauropsid reptile, appeared in the Pennsylvanian.*

Figure 100: *Petrolacosaurus, the first diapsid reptile known, lived during the late Carboniferous.*

Figure 101: *Archaeothyris was a very early synapsid and the oldest known.*

Fungi

Because plants and animals were growing in size and abundance in this time (for example, *Lepidodendron*), land fungi diversified further. Marine fungi still occupied the oceans. All modern classes of fungi were present in the Late Carboniferous (Pennsylvanian Epoch).[225]

Extinction events

Romer's gap

The first 15 million years of the Carboniferous had very limited terrestrial fossils. This gap in the fossil record is called Romer's gap after the American palaentologist Alfred Romer. While it has long been debated whether the gap is a result of fossilisation or relates to an actual event, recent work indicates the gap period saw a drop in atmospheric oxygen levels, indicating some sort of ecological collapse.[226] The gap saw the demise of the Devonian fish-like ichthyostegalian labyrinthodonts, and the rise of the more advanced temnospondyl and reptiliomorphan amphibians that so typify the Carboniferous terrestrial vertebrate fauna.

Carboniferous rainforest collapse

Before the end of the Carboniferous Period, an extinction event occurred. On land this event is referred to as the Carboniferous Rainforest Collapse (CRC). Vast tropical rainforests collapsed suddenly as the climate changed from hot and humid to cool and arid. This was likely caused by intense glaciation and a drop in sea levels.

The new climatic conditions were not favorable to the growth of rainforest and the animals within them. Rainforests shrank into isolated islands, surrounded by seasonally dry habitats. Towering lycopsid forests with a heterogeneous mixture of vegetation were replaced by much less diverse tree-fern dominated flora.

Amphibians, the dominant vertebrates at the time, fared poorly through this event with large losses in biodiversity; reptiles continued to diversify due to key adaptations that let them survive in the drier habitat, specifically the hard-shelled egg and scales, both of which retain water better than their amphibian counterparts.

Sources

- Dudley, Robert (1998). "Atmospheric Oxygen, Giant Paleozoic Insects and the Evolution of Aerial Locomotor Performance"[227]. *Journal of Experimental Biology.* **201**: 1043–1050.
- Menning, M.; Alekseev, A.S.; Chuvashov, B.I.; Davydov, V.I.; Devuyst, F.-X.; Forke, H.C.; Grunt, T.A.; Hance, L.; Heckel, P.H.; Izokh, N.G.; Jin, Y.-G.; Jones, P.J.; Kotlyar, G.V.; Kozur, H.W.; Nemyrovska, T.I.; Schneider, J.W.; Wang, X.-D.; Weddige, K.; Weyer, D. & Work, D.M. (2006). "Global time scale and regional stratigraphic reference scales of Central and West Europe, East Europe, Tethys, South China, and North America as used in the Devonian–Carboniferous–Permian Correlation Chart 2003 (DCP 2003)". *Palaeogeography, Palaeoclimatology, Palaeoecology.* **240** (1-2): 318–372. doi: 10.1016/j.palaeo.2006.03.058[228].
- Ogg, Jim (June 2004). "Overview of Global Boundary Stratotype Sections and Points (GSSP's)"[229]. Archived from the original[230] on April 23, 2006. Retrieved April 30, 2006.
- Stanley, S.M. (1999). *Earth System History.* New York: W.H. Freeman and Company. ISBN 0-7167-2882-6.

- ⊚ This article incorporates text from a publication now in the public domain: Chisholm, Hugh, ed. (1911). "Carboniferous System". *Encyclopædia Britannica* (11th ed.). Cambridge University Press.

External links

	Wikisource has original works on the topic: *Paleozoic#Carboniferous*

	Wikimedia Commons has media related to *Carboniferous*.

- "Geologic Time Scale 2004"[231]. International Commission on Stratigraphy (ICS). Archived from the original[232] on January 6, 2013. Retrieved January 15, 2013.
- Examples of Carboniferous Fossils[233]
- 60+ images of Carboniferous Foraminifera[234]

Permian

Permian Period *298.9–251.902 million years ago* PreЄЄ OSD C P T J K PgN	
Mean atmospheric O $_2$ content over period duration	c. 23 vol %[235,236] (115 % of modern level
Mean atmospheric CO $_2$ content over period duration	c. 900 ppm[237] (3 times pre-industrial level)
Mean surface temperature over period duration	c. 16 °C[238] (2 °C above modern level)
Sea level (above present day)	Relatively constant at 60 m (200 ft) in early Permian; plummeting during the middle Permian to a constant –20 m (–66 ft) in the late Permian.

Key events in the Permian

250
260
270
280
290
300

An approximate timescale of key Permian events.
Axis scale: millions of years ago.

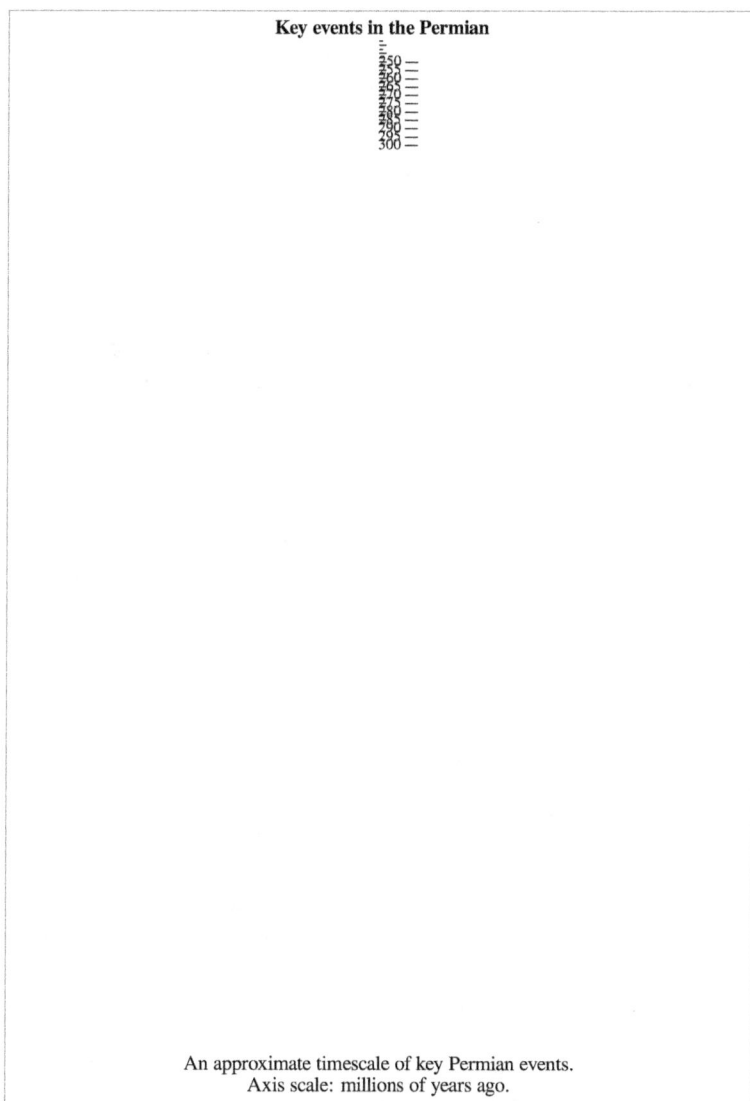

The **Permian** is a geologic period and system which spans 47 million years from the end of the Carboniferous Period 298.9 million years ago (Mya), to the beginning of the Triassic period 251.902 Mya. It is the last period of the Paleozoic era; the following Triassic period belongs to the Mesozoic era. The concept of the Permian was introduced in 1841 by geologist Sir Roderick Murchison, who named it after the city of Perm.

The Permian witnessed the diversification of the early amniotes into the ancestral groups of the mammals, turtles, lepidosaurs, and archosaurs. The world at the time was dominated by two continents known as Pangaea and Siberia, surrounded by a global ocean called Panthalassa. The Carboniferous rainforest collapse left behind vast regions of desert within the continental interior. Amniotes, who could better cope with these drier conditions, rose to dominance in place of their amphibian ancestors.

The Permian (along with the Paleozoic) ended with the Permian–Triassic extinction event, the largest mass extinction in Earth's history, in which nearly 90% of marine species and 70% of terrestrial species died out.[239] It would take well into the Triassic for life to recover from this catastrophe. Recovery from the Permian–Triassic extinction event was protracted; on land, ecosystems took 30 million years to recover.

Discovery

The term "Permian" was introduced into geology in 1841 by Sir R. I. Murchison, president of the Geological Society of London, who identified typical strata in extensive Russian explorations undertaken with Édouard de Verneuil.[240,241] The region now lies in the Perm Krai of Russia.

ICS Subdivisions

Official ICS 2017 subdivisions of the Permian System from most recent to most ancient rock layers are:

Lopingian epoch [259.8 ± 0.4 Mya - 251.902 ± 0.06 Mya]

- Changhsingian (Changxingian) [254.1 ± 0.07 Mya - 251.902 ± 0.06 Mya]
- Wuchiapingian (Wujiapingian) [259.8 ± 0.4 Mya - 254.1 ± 0.07 Mya]
- Others:
  - Waiitian (New Zealand) [260.4 ± 0.7 Mya - 253.8 ± 0.7 Mya]
  - Makabewan (New Zealand) [253.8 - 251.0 ± 0.4 Mya]
  - Ochoan (North American) [260.4 ± 0.7 Mya - 251.0 ± 0.4 Mya]

Guadalupian epoch [272.3 ± 0.5 - 259.8 ± 0.4 Mya]

- Capitanian stage [265.1 ± 0.4 - 259.8 ± 0.4 Mya]
- Wordian stage [268.8 ± 0.5 - 265.1 ± 0.4 Mya]
- Roadian stage [272.3 ± 0.5 - 268.8 ± 0.5 Mya]
- Others:
  - Kazanian or Maokovian (European) [270.6 ± 0.7 - 260.4 ± 0.7 Mya][242]
  - Braxtonian stage (New Zealand) [270.6 ± 0.7 - 260.4 ± 0.7 Mya]

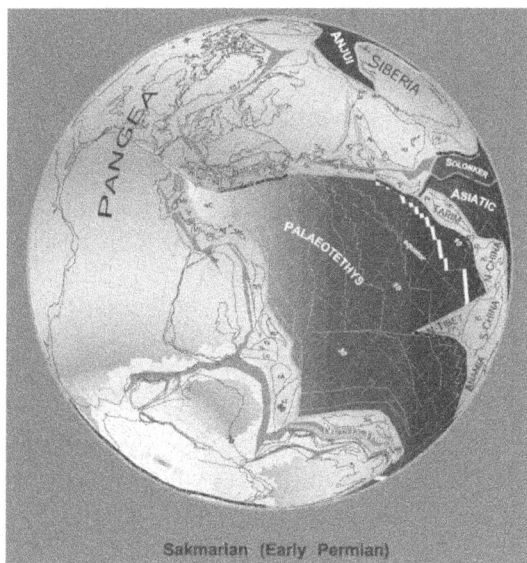

Figure 102: *Geography of the Permian world*

Cisuralian epoch [298.9 ± 0.15 - 272.3 ± 0.5 Mya]

- Kungurian stage [283.5 ± 0.7 - 272.3 ± 0.5 Mya]
- Artinskian stage [290.1 ± 0.7 - 283.5 ± 0.7 Mya]
- Sakmarian stage [295. ± 0.8 - 290.1 ± 0.7 Mya]
- Asselian stage [298.9 ± 0.15 - 294.6 ± 0.8 Mya]
- Others:
  - Telfordian (New Zealand) [289 - 278]
  - Mangapirian (New Zealand) [278 - 270.6]

Oceans

Sea levels in the Permian remained generally low, and near-shore environments were reduced as almost all major landmasses collected into a single continent – Pangaea. This could have in part caused the widespread extinctions of marine species at the end of the period by severely reducing shallow coastal areas preferred by many marine organisms.

Paleogeography

During the Permian, all the Earth's major landmasses were collected into a single supercontinent known as Pangaea. Pangaea straddled the equator and extended toward the poles, with a corresponding effect on ocean currents in the single great ocean ("Panthalassa", the "universal sea"), and the Paleo-Tethys Ocean, a large ocean that was between Asia and Gondwana. The Cimmeria continent rifted away from Gondwana and drifted north to Laurasia, causing the Paleo-Tethys Ocean to shrink. A new ocean was growing on its southern end, the Tethys Ocean, an ocean that would dominate much of the Mesozoic era. Large continental landmass interiors experience climates with extreme variations of heat and cold ("continental climate") and monsoon conditions with highly seasonal rainfall patterns. Deserts seem to have been widespread on Pangaea. Such dry conditions favored gymnosperms, plants with seeds enclosed in a protective cover, over plants such as ferns that disperse spores in a wetter environment. The first modern trees (conifers, ginkgos and cycads) appeared in the Permian.

Three general areas are especially noted for their extensive Permian deposits – the Ural Mountains (where Perm itself is located), China, and the southwest of North America, including the Texas red beds. The Permian Basin in the U.S. states of Texas and New Mexico is so named because it has one of the thickest deposits of Permian rocks in the world.

Climate

The climate in the Permian was quite varied. At the start of the Permian, the Earth was still in an ice age, which began in the Carboniferous. Glaciers receded around the mid-Permian period as the climate gradually warmed, drying the continent's interiors.[243] In the late Permian period, the drying continued although the temperature cycled between warm and cool cycles.

Life

Marine biota

Permian marine deposits are rich in fossil mollusks, echinoderms, and brachiopods. Fossilized shells of two kinds of invertebrates are widely used to identify Permian strata and correlate them between sites: fusulinids, a kind of shelled amoeba-like protist that is one of the foraminiferans, and ammonoids, shelled cephalopods that are distant relatives of the modern nautilus. By the close of the Permian, trilobites and a host of other marine groups became extinct.

Figure 103: *Selwyn Rock, South Australia, an exhumed glacial pavement of Permian age*

Figure 104: *Hercosestria cribrosa, a reef-forming productid brachiopod (Middle Permian, Glass Mountains, Texas).*

Terrestrial biota

Terrestrial life in the Permian included diverse plants, fungi, arthropods, and various types of tetrapods. The period saw a massive desert covering the interior of Pangaea. The warm zone spread in the northern hemisphere, where extensive dry desert appeared. The rocks formed at that time were stained red by iron oxides, the result of intense heating by the sun of a surface devoid of vegetation cover. A number of older types of plants and animals died out or became marginal elements.

The Permian began with the Carboniferous flora still flourishing. About the middle of the Permian a major transition in vegetation began. The swamp-loving lycopod trees of the Carboniferous, such as *Lepidodendron* and *Sigillaria*, were progressively replaced in the continental interior by the more advanced seed ferns and early conifers. At the close of the Permian, lycopod and equisete swamps reminiscent of Carboniferous flora survived only on a series of equatorial islands in the Paleo-Tethys Ocean that later would become South China.[244]

The Permian saw the radiation of many important conifer groups, including the ancestors of many present-day families. Rich forests were present in many areas, with a diverse mix of plant groups. The southern continent saw extensive seed fern forests of the *Glossopteris* flora. Oxygen levels were probably high there. The ginkgos and cycads also appeared during this period.

Insects

From the Pennsylvanian subperiod of the Carboniferous period until well into the Permian, the most successful insects were primitive relatives of cockroaches. Six fast legs, four well-developed folding wings, fairly good eyes, long, well-developed antennae (olfactory), an omnivorous digestive system, a receptacle for storing sperm, a chitin-based exoskeleton that could support and protect, as well as a form of gizzard and efficient mouth parts, gave it formidable advantages over other herbivorous animals. About 90% of insects at the start of the Permian were cockroach-like insects ("Blattopterans").[245]

Primitive forms of dragonflies (Odonata) were the dominant aerial predators and probably dominated terrestrial insect predation as well. True Odonata appeared in the Permian,[246,247] and all are effectively semi-aquatic insects (aquatic immature stages, and terrestrial adults), as are all modern odonates. Their prototypes are the oldest winged fossils,[248] dating back to the Devonian, and are different in several respects from the wings of other insects.[249] Fossils suggest they may have possessed many modern attributes even by the late Carboniferous, and it is possible that they captured small vertebrates, for at least one species had a wing span of 71 cm (28 in).[250] Several other insect groups appeared or flourished during the Permian, including the Coleoptera (beetles) and Hemiptera (true bugs).

Synapsid and amphibian fauna

Early Permian terrestrial faunas were dominated by pelycosaurs, diadectids and amphibians,[251] the middle Permian by primitive therapsids such as the dinocephalia, and the late Permian by more advanced therapsids such as gorgonopsians and dicynodonts. Towards the very end of the Permian the first archosaurs appeared, a group that would give rise to the crurotarsans and the dinosaurs in the following period. Also appearing at the end of the Permian were the first cynodonts, which would go on to evolve into mammals during the Triassic. Another group of therapsids, the therocephalians (such as *Lycosuchus*), arose in the Middle Permian. There were no aerial vertebrates (with the exception of gliding reptiles, the avicephalans).

The Permian period saw the development of a fully terrestrial fauna and the appearance of the first large herbivores and carnivores. It was the high tide of the anapsids in the form of the massive Pareiasaurs and host of smaller, generally lizard-like groups. A group of small reptiles, the diapsids, started to abound. These were the ancestors to most modern reptiles and the ruling dinosaurs as well as pterosaurs and crocodiles.

The synapsid, early ancestors to mammals, also thrived at this time. Synapsids included some large members such as *Dimetrodon*. The special adaptations of reptiles enabled them to flourish in the drier climate of the Permian and they grew to dominate the vertebrates.

Permian amphibians consisted of temnospondyli, lepospondyli and batrachosaurs.

Figure 105: *Edaphosaurus pogonias and Platyhystrix – Early Permian, North America and Europe*

Figure 106: *Dimetrodon grandis and Eryops – Early Permian, North America*

Figure 107: *Ocher fauna, Estemmenosuchus uralensis and Eotitanosuchus – Middle Permian, Ural Region*

Figure 108: *Titanophoneus and Ulemosaurus – Ural Region*

Figure 109: *Inostrancevia alexandri and Scutosaurus –*
Late Permian, North European Russia (Northern Dvina)

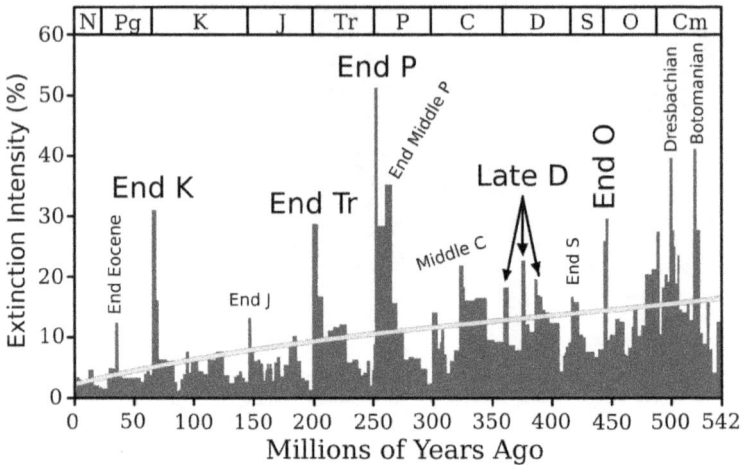

Figure 110: *The Permian–Triassic extinction event, labeled "End P" here, is the most significant extinction event in this plot for marine genera which produce large numbers of fossils.*

Permian–Triassic extinction event

The Permian ended with the most extensive extinction event recorded in paleontology: the Permian–Triassic extinction event. 90% to 95% of marine species became extinct, as well as 70% of all land organisms. It is also the only known mass extinction of insects. Recovery from the Permian-Triassic extinction event was protracted; on land, ecosystems took 30 million years to recover. Trilobites, which had thrived since Cambrian times, finally became extinct before the end of the Permian. Nautiluses, a species of cephalopods, surprisingly survived this occurrence.

There is evidence that magma, in the form of flood basalt, poured onto the surface in what is now called the Siberian Traps, for thousands of years, contributing to the environmental stress that led to mass extinction. The reduced coastal habitat and highly increased aridity probably also contributed. Based on the amount of lava estimated to have been produced during this period, the worst-case scenario is the release of enough carbon dioxide from the eruptions to raise world temperatures five degrees Celsius.

Another hypothesis involves ocean venting of hydrogen sulfide gas. Portions of the deep ocean will periodically lose all of its dissolved oxygen allowing bacteria that live without oxygen to flourish and produce hydrogen sulfide gas. If enough hydrogen sulfide accumulates in an anoxic zone, the gas can rise into the atmosphere. Oxidizing gases in the atmosphere would destroy the toxic gas, but the hydrogen sulfide would soon consume all of the atmospheric gas available. Hydrogen sulfide levels might have increased dramatically over a few hundred years. Models of such an event indicate that the gas would destroy ozone in the upper atmosphere allowing ultraviolet radiation to kill off species that had survived the toxic gas. There are species that can metabolize hydrogen sulfide.

Another hypothesis builds on the flood basalt eruption theory. An increase in temperature of five degrees Celsius would not be enough to explain the death of 95% of life. But such warming could slowly raise ocean temperatures until frozen methane reservoirs below the ocean floor near coastlines melted, expelling enough methane (among the most potent greenhouse gases) into the atmosphere to raise world temperatures an additional five degrees Celsius. The frozen methane hypothesis helps explain the increase in carbon-12 levels found midway in the Permian–Triassic boundary layer. It also helps explain why the first phase of the layer's extinctions was land-based, the second was marine-based (and starting right after the increase in C-12 levels), and the third land-based again.

An even more speculative hypothesis is that intense radiation from a nearby supernova was responsible for the extinctions.

It has been hypothesised that huge meteorite impact crater (Wilkes Land crater) with a diameter of around 500 kilometers in Antarctica represents an impact event that may be related to the extinction. The crater is located at a depth of 1.6 kilometers beneath the ice of Wilkes Land in eastern Antarctica. The scientists speculate that this impact may have caused the Permian–Triassic extinction event, although its age is bracketed only between 100 million and 500 million years ago. They also speculate that it may have contributed in some way to the separation of Australia from the Antarctic landmass, which were both part of a supercontinent called Gondwana. Levels of iridium and quartz fracturing in the Permian-Triassic layer do not approach those of the Cretaceous–Paleogene boundary layer. Given that a far greater proportion of species and individual organisms became extinct during the former, doubt is cast on the significance of a meteorite impact in creating the latter. Further doubt has been cast on this theory based on fossils in Greenland that show the extinction to have been gradual, lasting about eighty thousand years, with three distinct phases.

Many scientists argue that the Permian–Triassic extinction event was caused by a combination of some or all of the hypotheses above and other factors; the formation of Pangaea decreased the number of coastal habitats and may have contributed to the extinction of many clades.Wikipedia:Citation needed

Further reading

- Ogg, Jim (June 2004). "Overview of Global Boundary Stratotype Sections and Points (GSSP's)"[252]. *stratigraphy.org*. Archived from the original[253] on 2004-02-19. Retrieved April 30, 2006.

External links

Wikisource has original works on the topic: *Paleozoic#Permian*

Wikimedia Commons has media related to *Permian*.

- University of California offers a more modern Permian stratigraphy[254]
- Classic Permian strata in the Glass Mountains of the Permian Basin[255]
- "International Commission on Stratigraphy (ICS)"[256]. *Geologic Time Scale 2004*. Retrieved September 19, 2005.
- Examples of Permian Fossils[257]
- Fossil of Giant Permian Amphibian[258]
- Schneebeli-Hermann, Elke, "Extinguishing a Permian World"[259], *Geology*, **40** (3): 287–288, Bibcode: 2012Geo....40..287S[260], doi: 10.1130/focus032012.1[261]

Mesozoic

Mesozoic Era
251.902 - 66 million years ago

Key events in the Mesozoic

Phanerozoic
An approximate timescale of key Mesozoic events.
Axis scale: millions of years ago.

The **Mesozoic Era** (/ˌmɛsəˈzoʊɪk, <wbr />ˌmiː-, <wbr />-soʊ-/ or /ˌmɛzəˈzoʊɪk, <wbr />ˌmiː-, <wbr />-soʊ-/[262,263]) is an interval of geological time from about 252 to 66[264] million years ago. It is also called the **Age of Reptiles** a phrase introduced by the 19th century paleontologist Gideon Mantell who viewed it as dominated by diapsids such as *Iguanodon*, *Megalosaurus*, *Plesiosaurus* and *Pterodactylus*. To paleobotanists, this Era is also called the **Age of Conifers**.

Mesozoic means "middle life", deriving from the Greek prefix *meso-/μεσο-* for "between" and *zōon/ζῷον* meaning "animal" or "living being". The name "Mesozoic" was proposed in 1840 by the British geologist John Phillips (1800–1874). It is one of three geologic eras of the Phanerozoic Eon, preceded by the Paleozoic ("ancient life") and succeeded by the Cenozoic ("new life"). The era is subdivided into three major periods: the Triassic, Jurassic, and Cretaceous, which are further subdivided into a number of epochs and stages.

The era began in the wake of the Permian–Triassic extinction event, the largest well-documented mass extinction in Earth's history, and ended with the Cretaceous–Paleogene extinction event, another mass extinction whose victims included the non-avian dinosaurs. The Mesozoic was a time of significant tectonic, climate and evolutionary activity. The era witnessed the gradual rifting of the supercontinent Pangaea into separate landmasses that would move into their current positions during the next era. The climate of the Mesozoic was varied, alternating between warming and cooling periods. Overall, however, the Earth was hotter than it is today. Dinosaurs appeared in the Late Triassic and became the dominant terrestrial vertebrates early in the Jurassic, occupying this position for about 135 million years until their demise at the end of the Cretaceous. Birds first appeared in the Jurassic, having evolved from a branch of theropod dinosaurs. The first mammals also appeared during the Mesozoic, but would remain small—less than 15 kg (33 lb)—until the Cenozoic.

Geologic periods

Following the Paleozoic, the Mesozoic extended roughly 186 million years, from 251.902 to 66[264] million years ago when the Cenozoic Era began. This time frame is separated into three geologic periods. From oldest to youngest:

- Triassic (251.902 to 201.3[265] million years ago)
- Jurassic (201.3 to 145[266] million years ago)
- Cretaceous (145 to 66[267] million years ago)

The lower boundary of the Mesozoic is set by the Permian–Triassic extinction event, during which approximately 90% to 96% of marine species and 70% of terrestrial vertebrates became extinct. It is also known as the "Great Dying" because it is considered the largest mass extinction in the Earth's history. The upper boundary of the Mesozoic is set at the Cretaceous–Paleogene extinction event (or K–Pg extinction event), which may have been caused by an asteroid impactor that created Chicxulub Crater on the Yucatán Peninsula. Towards the Late Cretaceous, large volcanic eruptions are also believed to have contributed to the Cretaceous–Paleogene extinction event. Approximately 50% of all genera became extinct, including all of the non-avian dinosaurs.

Triassic

The Triassic ranges roughly from 252 million to 201 million years ago, preceding the Jurassic Period. The period is bracketed between the Permian–Triassic extinction event and the Triassic–Jurassic extinction event, two of the "big five", and it is divided into three major epochs: Early, Middle, and Late Triassic.

The Early Triassic, about 252 to 247 million years ago, was dominated by deserts in the interior of the Pangaea supercontinent. The Earth had just witnessed a massive die-off in which 95% of all life became extinct, and the most

Figure 111: *Plateosaurus (a prosauropod)*

common vertebrate life on land were *lystrosaurus*, labyrinthodonts, and *euparkeria* along with many other creatures that managed to survive the Permian extinction. Temnospondyls evolved during this time and would be the dominant predator for much of the Triassic.

The Middle Triassic, from 247 to 237 million years ago, featured the beginnings of the breakup of Pangaea and the opening of the Tethys Sea. Ecosystems had recovered from the Permian extinction. Algae, sponge, corals, and crustaceans all had recovered, and new aquatic reptiles evolved, such as ichthyosaurs and nothosaurs. On land, pine forests flourished, as did groups of insects like mosquitoes and fruit flies. Reptiles began to get bigger and bigger, and the first crocodilians evolved, which sparked competition with the large amphibians that had previously ruled the freshwater world.

Following the bloom of the Middle Triassic, the Late Triassic, from 237 to 201 million years ago, featured frequent heat spells and moderate precipitation (10-20 inches per year). The recent warming led to a boom of reptilian evolution on land as the first true dinosaurs evolved, as well as pterosaurs. During the Late Triassic, some advanced cynodonts gave rise to the first Mammaliaformes. All this climatic change, however, resulted in a large die-out known as the Triassic-Jurassic extinction event, in which many archosaurs (excluding pterosaurs, dinosaurs and crocodylomorphs), most synapsids, and almost all large amphibians became extinct, as well as 34% of marine life, in the Earth's fourth mass extinction event. The cause is debatable;flood basalt eruptions at the Central Atlantic magmatic province is cited as one possible cause.

Figure 112: *Rhamphorhynchus*

Jurassic

The Jurassic ranges from 200 million years to 145 million years ago and features 3 major epochs: The Early Jurassic, the Middle Jurassic, and the Late Jurassic.

The Early Jurassic spans from 200 to 175 million years ago. The climate was tropical, much more humid than the Triassic. In the oceans, plesiosaurs, ichthyosaurs and ammonites were abundant. On land, dinosaurs and other archosaurs staked their claim as the dominant race, with species such as *Dilophosaurus* at the top of the food chain. The first true crocodiles evolved, pushing the large amphibians to near extinction. All-in-all, archosaurs rose to rule the world. Meanwhile, the first true mammals evolved, remaining relatively small but spreading widely; the Jurassic *Castorocauda*, for example, had adaptations for swimming, digging and catching fish. *Fruitafossor*, from the late Jurassic period about 150 million years ago, was about the size of a chipmunk, and its teeth, forelimbs and back suggest that it dug open the nests of social insects (probably termites, as ants had not yet appeared). The first multituberculates like *Rugosodon* evolved, while volaticotherians took to the skies.

The Middle Jurassic spans from 175 to 163 million years ago. During this epoch, dinosaurs flourished as huge herds of sauropods, such as *Brachiosaurus* and *Diplodocus*, filled the fern prairies, chased by many new predators such as *Allosaurus*. Conifer forests made up a large portion of the forests. In the oceans, plesiosaurs were quite common, and ichthyosaurs flourished. This epoch was the peak of the reptiles.

Figure 113: *Stegosaurus*

The Late Jurassic spans from 163 to 145 million years ago. During this epoch, the first avialans, like *Archaeopteryx*, evolved from small coelurosaurian dinosaurs. The increase in sea levels opened up the Atlantic seaway, which has grown continually larger until today. The divided landmasses gave opportunity for the diversification of new dinosaurs.

Cretaceous

The Cretaceous is the longest period of the Mesozoic, but has only two epochs: Early and Late Cretaceous.

The Early Cretaceous spans from 145 to 100 million years ago. The Early Cretaceous saw the expansion of seaways, and as a result, the decline and extinction of sauropods (except in South America). Some island-hopping dinosaurs, like *Eustreptospondylus*, evolved to cope with the coastal shallows and small islands of ancient Europe. Other dinosaurs rose up to fill the empty space that the Jurassic-Cretaceous extinction left behind, such as *Carcharodontosaurus* and *Spinosaurus*. Of the most successful was the *Iguanodon*, which spread to every continent. Seasons came back into effect and the poles got seasonally colder, but some dinosaurs still inhabited the polar forests year round, such as *Leaellynasaura* and *Muttaburrasaurus*. The poles were too cold for

Figure 114: *Tylosaurus (a mosasaur) hunting Xiphactinus*

crocodiles, and became the last stronghold for large amphibians like *Koola-suchus*. Pterosaurs got larger as species like *Tapejara* and *Ornithocheirus* evolved. Mammals continued to expand their range: eutriconodonts produced fairly large, wolverine-like predators like *Repenomamus* and *Gobicon-odon*, early therians began to expand into metatherians and eutherians, and cimolodont multituberculates went on to become common in the fossil record.

The Late Cretaceous spans from 100 to 66 million years ago. The Late Creta-ceous featured a cooling trend that would continue in the Cenozoic era. Even-tually, tropics were restricted to the equator and areas beyond the tropic lines experienced extreme seasonal changes in weather. Dinosaurs still thrived, as new species such as *Tyrannosaurus, Ankylosaurus, Triceratops* and hadrosaurs dominated the food web. In the oceans, mosasaurs ruled, filling the role of the ichthyosaurs, which, after declining, had disappeared in the Cenomanian-Turonian boundary event. Though pliosaurs had gone extinct in the same event, long-necked plesiosaurs such as *Elasmosaurus* continued to thrive. Flowering plants, possibly appearing as far back as the Triassic, became truly dominant for the first time. Pterosaurs in the Late Cretaceous declined for poorly understood reasons, though this might be due to tendencies of the fos-sil record, as their diversity seems to be much higher than previously thought. Birds became increasingly common and diversified into a variety of enantior-nithe and ornithurine forms. Though mostly small, marine hesperornithes be-came relatively large and flightless, adapted to life in the open sea. Metathe-

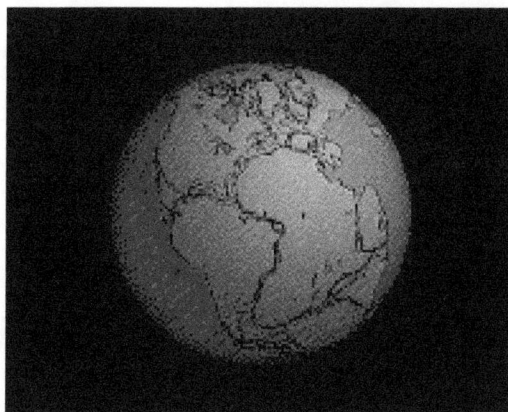

Figure 115: *Breakup of Pangaea*

rians and primitive eutherian also became common and even produced large and specialised species like *Didelphodon* and *Schowalteria*. Still, the dominant mammals were multituberculates, cimolodonts in the north and gondwanatheres in the south. At the end of the Cretaceous, the Deccan traps and other volcanic eruptions were poisoning the atmosphere. As this continued, it is thought that a large meteor smashed into earth 66 million years ago, creating the Chicxulub Crater in an event known as the K-Pg Extinction (formerly K-T), the fifth and most recent mass extinction event, in which 75% of life became extinct, including all non-avian dinosaurs. Everything over 10 kilograms became extinct. The age of the dinosaurs was over.

Paleogeography and tectonics

Compared to the vigorous convergent plate mountain-building of the late Paleozoic, Mesozoic tectonic deformation was comparatively mild. The sole major Mesozoic orogeny occurred in what is now the Arctic, creating the Innuitian orogeny, the Brooks Range, the Verkhoyansk and Cherskiy Ranges in Siberia, and the Khingan Mountains in Manchuria. This orogeny was related to the opening of the Arctic Ocean and subduction of the North China and Siberian cratons under the Pacific Ocean.[268] In contrast, the era featured the dramatic rifting of the supercontinent Pangaea, which gradually split into a northern continent, Laurasia, and a southern continent, Gondwana. This created the passive continental margin that characterizes most of the Atlantic coastline (such as along the U.S. East Coast) today.[269]

By the end of the era, the continents had rifted into nearly their present forms, though not their present positions. Laurasia became North America and Eurasia, while Gondwana split into South America, Africa, Australia, Antarctica and the Indian subcontinent, which collided with the Asian plate during the Cenozoic, giving rise to the Himalayas.

Climate

The Triassic was generally dry, a trend that began in the late Carboniferous, and highly seasonal, especially in the interior of Pangaea. Low sea levels may have also exacerbated temperature extremes. With its high specific heat capacity, water acts as a temperature-stabilizing heat reservoir, and land areas near large bodies of water—especially oceans—experience less variation in temperature. Because much of Pangaea's land was distant from its shores, temperatures fluctuated greatly, and the interior probably included expansive deserts. Abundant red beds and evaporites such as halite support these conclusions, but some evidence suggests the generally dry climate of was punctuated by episodes of increased rainfall. The most important humid episodes were the Carnian Pluvial Event and one in the Rhaetian, a few million years before the Triassic–Jurassic extinction event.

Sea levels began to rise during the Jurassic, probably caused by an increase in seafloor spreading. The formation of new crust beneath the surface displaced ocean waters by as much as 200 m (656 ft) above today's sea level, flooding coastal areas. Furthermore, Pangaea began to rift into smaller divisions, creating new shoreline along the Tethys Sea. Temperatures continued to increase, then began to stabilize. Humidity also increased with the proximity of water, and deserts retreated.

The climate of the Cretaceous is less certain and more widely disputed. Probably, higher levels of carbon dioxide in the atmosphere are thought to have almost eliminated the north-south temperature gradient: temperatures were about the same across the planet, and about 10°C higher than today. The circulation of oxygen to the deep ocean may also have been disrupted,[270]Wikipedia:Accuracy dispute#Disputed statement preventing the decomposition of large volumes of organic matter, which was eventually deposited as "black shale".

Not all data support these hypotheses, however. Even with the overall warmth, temperature fluctuations should have been sufficient for the presence of polar ice caps and glaciers, but there is no evidence of either. Quantitative models have also been unable to recreate the flatness of the Cretaceous temperature gradient.Wikipedia:Citation needed

Figure 116: *Conifers were the dominant terrestrial plants for most of the Mesozoic. Flowering plants appeared late in the era but did not become widespread until the Cenozoic.*

Different studies have come to different conclusions about the amount of oxygen in the atmosphere during different parts of the Mesozoic, with some concluding oxygen levels were lower than the current level (about 21%) throughout the Mesozoic,[271,272] some concluding they were lower in the Triassic and part of the Jurassic but higher in the Cretaceous,[273,274,275] and some concluding they were higher throughout most or all of the Triassic, Jurassic and Cretaceous.[276,277]

Life

Flora

The dominant land plant species of the time were gymnosperms, which are vascular, cone-bearing, non-flowering plants such as conifers that produce seeds without a coating. This is opposed to the earth's current flora, in which the dominant land plants in terms of number of species are angiosperms. One particular plant genus, *Ginkgo*, is thought to have evolved at this time and is represented today by a single species, *Ginkgo biloba*. As well, the extant genus *Sequoia* is believed to have evolved in the Mesozoic.[278]

Figure 117: *Dinosaurs were the dominant terres-trial vertebrates throughout much of the Mesozoic.*

Some plant species had distributions that were markedly different from suc-ceeding periods; for example, the Schizeales, a fern order, were skewed to the Northern Hemisphere in the Mesozoic, but are now better represented in the Southern Hemisphere.[279]

Fauna

The extinction of nearly all animal species at the end of the Permian Period allowed for the radiation of many new lifeforms. In particular, the extinction of the large herbivorous pareiasaurs and carnivorous gorgonopsians left those ecological niches empty. Some were filled by the surviving cynodonts and dicynodonts, the latter of which subsequently became extinct.

Recent research indicates that it took much longer for the reestablishment of complex ecosystems with high biodiversity, complex food webs, and spe-cialized animals in a variety of niches, beginning in the mid-Triassic 4M to 6M years after the extinction, and not complete until 30M years after the extinction. Animal life was then dominated by various archosaurs: di-nosaurs, pterosaurs, and aquatic reptiles such as ichthyosaurs, plesiosaurs, and mosasaurs.

The climatic changes of the late Jurassic and Cretaceous favored further adaptive radiation. The Jurassic was the height of archosaur diversity, and the first birds and eutherian mammals also appeared. Flowering plants radiated sometime in the early Cretaceous, first in the tropics, but the even temperature gradient allowed them to spread toward the poles throughout the period. By the end of the Cretaceous, angiosperms dominated tree floras in many areas, although some evidence suggests that biomass was still dominated by cycads and ferns until after the Cretaceous–Paleogene extinction.

Some have argued that insects diversified in symbiosis with angiosperms, because insect anatomy, especially the mouth parts, seems particularly well-suited for flowering plants. However, all major insect mouth parts preceded angiosperms, and insect diversification actually slowed when they arrived, so their anatomy originally must have been suited for some other purpose.

References

<templatestyles src="Template:Refbegin/styles.css" />

* *British Mesozoic Fossils*, 1983, The Natural History Museum, London.

External links

Wikisource has original works on the topic: *Mesozoic*

* 🔊 Media related to Mesozoic at Wikimedia Commons

Triassic

Triassic Period	
251.902–201.3 million years ago	
PreЄЄ OSD C P T J K PgN	
Mean atmospheric O $_2$ content over period duration	c. 16 vol %[280,281] (80 % of modern level
Mean atmospheric CO $_2$ content over period duration	c. 1750 ppm[282] (6 times pre-industrial level)
Mean surface temperature over period duration	c. 17 °C[283] (3 °C above modern level)

Key events in the Triassic

An approximate timescale of key Triassic events.
Axis scale: millions of years ago.

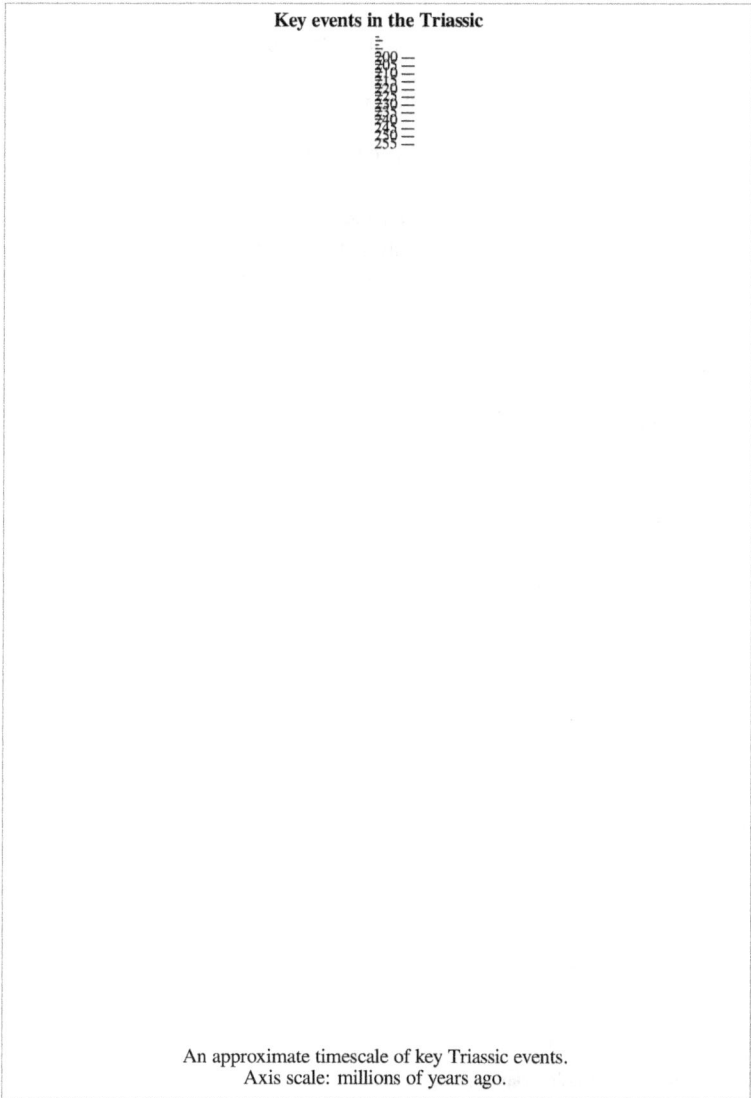

The **Triassic** (/traɪˈæsɪk/) is a geologic period and system which spans 50.6 million years from the end of the Permian Period 251.9 million years ago (Mya), to the beginning of the Jurassic Period 201.3 Mya. The Triassic is the first period of the Mesozoic Era. Both the start and end of the period are marked by major extinction events.

The Triassic began in the wake of the Permian–Triassic extinction event, which left the earth's biosphere impoverished; it would take well into the middle of this period for life to recover its former diversity. Therapsids and archosaurs were the chief terrestrial vertebrates during this time. A specialized subgroup of archosaurs, called dinosaurs, first appeared in the Late Triassic but did not become dominant until the succeeding Jurassic Period.

The first true mammals, themselves a specialized subgroup of Therapsids, also evolved during this period, as well as the first flying vertebrates, the pterosaurs, who like the dinosaurs were a specialized subgroup of archosaurs. The vast supercontinent of Pangaea existed until the mid-Triassic, after which it began to gradually rift into two separate landmasses, Laurasia to the north and Gondwana to the south.

The global climate during the Triassic was mostly hot and dry, with deserts spanning much of Pangaea's interior. However, the climate shifted and became more humid as Pangaea began to drift apart. The end of the period was marked by yet another major mass extinction, the Triassic–Jurassic extinction event, that wiped out many groups and allowed dinosaurs to assume dominance in the Jurassic.

The Triassic was named in 1834 by Friedrich von Alberti, after the three distinct rock layers (*tri* meaning "three") that are found throughout Germany and northwestern Europe—red beds, capped by marine limestone, followed by a series of terrestrial mud- and sandstones—called the "Trias".[284]

Dating and subdivisions

The Triassic is usually separated into Early, Middle, and Late Triassic Epochs, and the corresponding rocks are referred to as Lower, Middle, or Upper Triassic. The faunal stages from the youngest to oldest are:

Upper/Late Triassic (Tr3)	
Rhaetian	(208.5 – 201.3 Mya)
Norian	(227 – 208.5 Mya)
Carnian	(237– 227 Mya)
Middle Triassic (Tr2)	
Ladinian	(242 – 237 Mya)
Anisian	(247.2 – 242 Mya)
Lower/Early Triassic (Scythian)	
Olenekian	(251.2 – 247.2 Mya)
Induan	(251.902– 251.2 Mya)

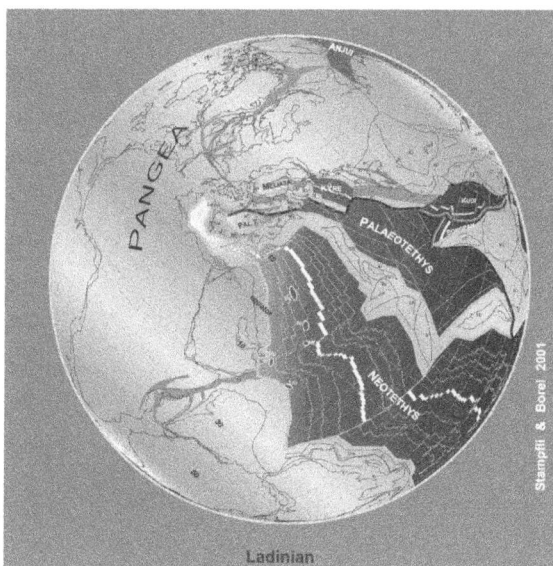

Figure 118: *230 Ma plate tectonic reconstruction*

Paleogeography

During the Triassic, almost all the Earth's land mass was concentrated into a single supercontinent centered more or less on the equator and spanning from pole to pole, called Pangaea ("all the land"). From the east, along the equator, the Tethys sea penetrated Pangaea, causing the Paleo-Tethys Ocean to be closed.

Later in the mid-Triassic a similar sea penetrated along the equator from the west. The remaining shores were surrounded by the world-ocean known as Panthalassa ("all the sea"). All the deep-ocean sediments laid down during the Triassic have disappeared through subduction of oceanic plates; thus, very little is known of the Triassic open ocean.

The supercontinent Pangaea was rifting during the Triassic—especially late in that period—but had not yet separated. The first nonmarine sediments in the rift that marks the initial break-up of Pangaea, which separated New Jersey from Morocco, are of Late Triassic age; in the U.S., these thick sediments comprise the Newark Group.

Because a super-continental mass has less shoreline compared to one broken up, Triassic marine deposits are globally relatively rare, despite their prominence in Western Europe, where the Triassic was first studied. In North America, for example, marine deposits are limited to a few exposures in the west.

Figure 119: *Sydney, Australia lies on Triassic shales and sandstones. Almost all of the exposed rocks around Sydney belong to the Triassic Sydney sandstone.*

Thus Triassic stratigraphy is mostly based on organisms that lived in lagoons and hypersaline environments, such as Estheria crustaceans.

Africa

At the beginning of the Mesozoic Era, Africa was joined with Earth's other continents in Pangaea. Africa shared the supercontinent's relatively uniform fauna which was dominated by theropods, prosauropods and primitive ornithischians by the close of the Triassic period. Late Triassic fossils are found throughout Africa, but are more common in the south than north. The time boundary separating the Permian and Triassic marks the advent of an extinction event with global impact, although African strata from this time period have not been thoroughly studied.[285]

Scandinavia

During the Triassic peneplains are thought to have formed in what is now Norway and southern Sweden. Remnants of this peneplain can be traced as a tilted summit accordance in the Swedish West Coast. In northern Norway Triassic peneplains may have been buried in sediments to be then re-exposed as coastal plains called strandflats. Dating of illite clay from a strandflat of Bømlo, southern Norway, have shown that landscape there became weathered in Late Triassic times (*c*. 210 million years ago) with the landscape likely also being shaped during that time.

South America

At Paleorrota geopark, located in Rio Grande do Sul, Brazil, the Santa Maria Formation and Caturrita Formations are exposed. In these formations, one of the earliest dinosaurs, *Staurikosaurus*, as well as the mammal ancestors *Brasilitherium* and *Brasilodon* have been discovered.

Climate

The Triassic continental interior climate was generally hot and dry, so that typical deposits are red bed sandstones and evaporites. There is no evidence of glaciation at or near either pole; in fact, the polar regions were apparently moist and temperate, providing a climate suitable for forests and vertebrates, including reptiles. Pangaea's large size limited the moderating effect of the global ocean; its continental climate was highly seasonal, with very hot summers and cold winters.[286] The strong contrast between the Pangea supercontinent and the global ocean triggered intense cross-equatorial monsoons.

The Triassic may have mostly been a dry period, but evidence exists that it was punctuated by several episodes of increased rainfall in tropical and subtropical latitudes of the Tethys Sea and its surrounding land. Sediments and fossils suggestive of a more humid climate are known from the Anisian to Ladinian of the Tethysian domain, and from the Carnian and Rhaetian of a larger area that includes also the Boreal domain (e.g., Svalbard Islands), the North American continent, the South China block and Argentina.

The best studied of such episodes of humid climate, and probably the most intense and widespread, was the Carnian Pluvial Event.

Life

Three categories of organisms can be distinguished in the Triassic record: holdovers from the Permian-Triassic extinction, new groups which flourished briefly, and other new groups which went on to dominate the Mesozoic Era.

Flora

On land, the surviving vascular plants included the lycophytes, the dominant cycadophytes, ginkgophyta (represented in modern times by *Ginkgo biloba*), ferns, horsetails and glossopterids. The spermatophytes, or seed plants, came to dominate the terrestrial flora: in the northern hemisphere, conifers, ferns and bennettitales flourished. *Glossopteris* (a seed fern) was the dominant southern hemisphere tree during the Early Triassic period.

Figure 120: *Triassic florá as depicted in Meyers Konversations-Lexikon (1885–90)*

Griesbachian-Dienerian-Smithian | Spathian-Anisian

Figure 121: *Marine vertebrate apex predators of the early Triassic*

Marine fauna

In marine environments, new modern types of corals appeared in the Early Triassic, forming small patches of reefs of modest extent compared to the great reef systems of Devonian or modern times. Serpulids appeared in the Middle Triassic. Microconchids were abundant. The shelled cephalopods called ammonites recovered, diversifying from a single line that survived the Permian extinction.

Figure 122: *Middle Triassic marginal marine sequence, southwestern Utah*

The fish fauna was remarkably uniform, suggesting that very few families survived the Permian extinction. There were also many types of marine reptiles. These included the Sauropterygia, which featured pachypleurosaurus and nothosaurs (both common during the Middle Triassic, especially in the Tethys region), placodonts, and the first plesiosaurs. The first of the lizardlike Thalattosauria (askeptosaurs) and the highly successful ichthyosaurs, which appeared in Early Triassic seas soon diversified, and some eventually developed to huge size during the late Triassic. Subequatorial saurichthyids have also been described in Early Triassic strata.

Terrestrial and freshwater fauna

Groups of terrestrial fauna, which appeared in the Triassic period or achieved a new level of evolutionary success during it include:

- Temnospondyls: one of the largest groups of early amphibians, temnospondyls originated during the Carboniferous and were still significant. Once abundant in both terrestrial and aquatic environments, the terrestrial species had mostly been replaced by reptiles. The Triassic survivors were aquatic or semi-aquatic, and were represented by *Tupilakosaurus*, *Thabanchuia*, Branchiosauridae and *Micropholis*, all of which died out in early Triassic, and the successful Stereospondyli, with survivors into the Cretaceous era. The largest of these, such as the *Mastodonsaurus* were up to 13 ft in length.
- Rhynchosaurs, barrel-gutted herbivores which thrived for only a short period of time, becoming extinct about 220 million years ago. They were exceptionally abundant in Triassic, the primary large herbivores in many

ecosystems. They sheared plants with their beaks and several rows of teeth on the roof of the mouth.

- Phytosaurs: archosaurs that prospered during the late Triassic. These long-snouted and semiaquatic predators resemble living crocodiles and probably had a similar lifestyle, hunting for fish and small reptiles around the water's edge. However this resemblance is only superficial and is a prime-case of convergent evolution.

- Aetosaurs: heavily armored archosaurs that were common during the last 30 million years of the late Triassic but died out at the Triassic-Jurassic extinction. Most aetosaurs were herbivorous, and fed on low-growing plants but some may have eaten meat.

- Rauisuchians, another group of archosaurs, which were the keystone predators of most Triassic terrestrial ecosystems. Over 25 species have been found, and include giant quadrupedal hunters, sleek bipedal omnivores, and lumbering beasts with deep sails on their backs. They probably occupied the large-predator niche later filled by theropods.

- Theropods: dinosaurs that first evolved in the Triassic period but did not evolve into large sizes until the Jurassic. Most Triassic theropods, such as the *Coelophysis*, were only around 1–2 meters long and hunted small prey in the shadow of the giant Rauisuchians.

- Cynodonts, a large group that includes true mammals. The first cynodonts evolved in the Permian, but many groups prospered during the Triassic. Their characteristic mammalian features included hair, a large brain, and upright posture. Many were small but several forms were enormous and filled a large herbivore niche before the evolution of sauropodomorph dinosaurs, as well as large sized carnivorous niches.

The Permian-Triassic extinction devastated terrestrial life. Biodiversity rebounded as the surviving species repopulated empty terrain, but these were short lived. Diverse communities with complex food-web structures took 30 million years to reestablish.

Temnospondyl amphibians were among those groups that survived the Permian-Triassic extinction; some lineages (e.g. trematosaurs) flourished briefly in the Early Triassic, while others (e.g. capitosaurs) remained successful throughout the whole period, or only came to prominence in the Late Triassic (e.g. plagiosaurs, metoposaurs). As for other amphibians, the first Lissamphibia, progenitors of first frogs, are known from the Early Triassic, but the group as a whole did not become common until the Jurassic, when the temnospondyls had become very rare. Other survivors the Chroniosuchia and Embolomeri were more closely related to amniotes than temnospondyls. Those became extinct after some million years.

Most of the Reptiliomorpha, stem-amniotes that gave rise to the amniotes, disappeared in Triassic, but two water dwelling groups survived; Embolomeri that only survived into the early part of the period, and the Chroniosuchia, which survived until the end of Triassic.

Archosauromorph reptiles, especially archosaurs, progressively replaced the synapsids that had dominated the previous Permian period. The *Cynognathus* was the characteristic top predator in earlier Triassic (Olenekian and Anisian) on Gondwana. Both kannemeyeriid dicynodonts and gomphodont cynodonts remained important herbivores during much of the period, and ecteniniids played a role as large sized, cursorial predators in the late Triassic. During the Carnian (early part of the Late Triassic), some advanced cynodonts gave rise to the first mammals. At the same time the Ornithodira, which until then had been small and insignificant, evolved into pterosaurs and a variety of dinosaurs. The Crurotarsi were the other important archosaur clade, and during the Late Triassic these also reached the height of their diversity, with various groups including the phytosaurs, aetosaurs, several distinct lineages of Rauisuchia, and the first crocodylians (the Sphenosuchia). Meanwhile, the stocky herbivorous rhynchosaurs and the small to medium-sized insectivorous or piscivorous Prolacertiformes were important basal archosauromorph groups throughout most of the Triassic.

Among other reptiles, the earliest turtles, like *Proganochelys* and *Proterochersis*, appeared during the Norian Age (Stage) of the late Triassic Period. The Lepidosauromorpha, specifically the Sphenodontia, are first found in the fossil record of the earlier Carnian Age. The Procolophonidae were an important group of small lizard-like herbivores.

During the Triassic, archosaurs displaced therapsids as the dominant amniotes. This "Triassic Takeover" may have contributed to the evolution of mammals by forcing the surviving therapsids and their mammaliaform successors to live as small, mainly nocturnal insectivores. Nocturnal life may have forced the mammaliaforms to develop fur and a higher metabolic rate.

Postosuchus kirkpatricki
50 cm

Figure 123: *Postosuchus An apex predator of its time which preyed on anything smaller than itself.*

Figure 124: *Staurikosaurus feeding on a dicynodont, in geopark Paleorrota.*

Figure 125: *Lystrosaurus was the most common land vertebrate during the Early Triassic, when animal life had been greatly diminished.*

Figure 126: *Reconstruction of Proterosuchus, a genus of crocodile-like carnivorous reptile that existed in the Early Triassic.*

Figure 127: *Cynognathus was a mammal-like cynodont from the early Triassic. The first true mammals evolved during this period.*

Figure 128: *Plateosaurus was an early sauropodomorph, or "prosauropod", of the late Triassic.*

Figure 129: *Coelophysis, one of the first dinosaurs, appeared in the Late Triassic.*

Figure 130: *Life reconstruction of Tanystropheus longobardicus*

Coal

- No known coal deposits date from the start of the Triassic period. This is known as the "coal gap" and can be seen as part of the Permian–Triassic extinction event. Sharp drops in sea level at the time of the Permo Triassic boundary may be the best explanation for the coal gap. However, there is still speculation as to why it is missing. During the preceding Permian period the arid desert conditions contributed to the evaporation of many inland seas and the subsequent inundation of those area, perhaps by a number of tsunami events that may have been responsible for the drop in sea level. Large salt basins in the southwest United States and a very large basin in central Canada are evidence of this.[287] Coal of Triassic age is made of the fossilized remains of Triassic plants and trees.

- Immediately above the Permian-Triassic boundary the glossopteris flora was suddenly largely displaced by an Australia-wide coniferous flora containing few species with a lycopod herbaceous under story. Conifers also became common in Eurasia. These groups of conifers arose from endemic species because of the ocean barriers that prevented seed from crossing for over one hundred million years. For instance, Podocarpis was located south and pines, junipers, and sequoias were located north. The dividing line ran through the Amazon Valley, across the Sahara, and north

Figure 131: *Triassic sandstone near Stadtroda, Germany*

of Arabia, India, Thailand, and Australia. It has been suggested that there was a climate barrier for the conifers.[288] although water barriers are more plausible. If so, something that could cross at least minor water barriers must have been involved in producing the coal hiatus. Hot climate could have been an important auxiliary factor across Antarctica or the Bering Strait, however. There was a spike of fern and lycopod spores immediately after the close of the Permian. There was also a spike of fungal spores immediately after the Permian-Triassic boundary. This spike may have lasted 50,000 years in Italy and 200,000 years in China and must have contributed to the climate warmth.

- An event excluding a catastrophe must have been involved to cause the coal hiatus because fungi would have removed all dead vegetation and coal forming detritus in a few decades in most tropical places. In addition, fungal spores rose gradually and declined similarly along with a prevalence of woody debris. Each phenomenon would hint at widespread vegetative death. Whatever the cause of the coal hiatus, it must have started in North America approximately 25 million years sooner.

Lagerstätten

The Monte San Giorgio lagerstätte, now in the Lake Lugano region of northern Italy and Switzerland, was in Triassic times a lagoon behind reefs with an

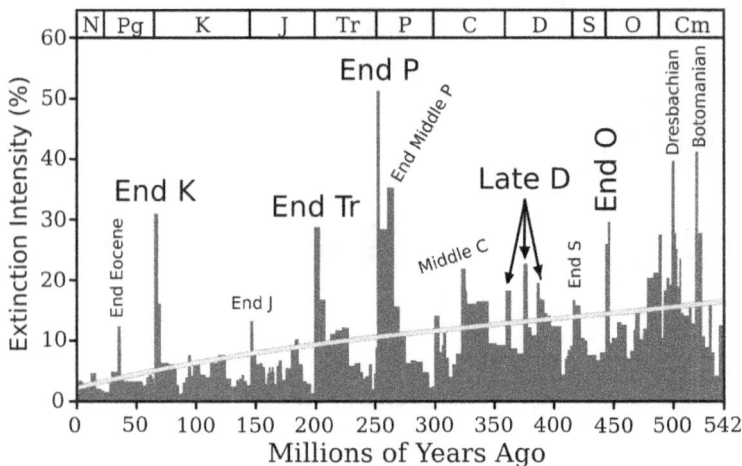

Figure 132: *The mass extinction event is marked by 'End Tr'*

anoxic bottom layer, so there were no scavengers and little turbulence to disturb fossilization, a situation that can be compared to the better-known Jurassic Solnhofen limestone lagerstätte.

The remains of fish and various marine reptiles (including the common pachypleurosaur Neusticosaurus, and the bizarre long-necked archosauromorph *Tanystropheus*), along with some terrestrial forms like *Ticinosuchus* and *Macrocnemus*, have been recovered from this locality. All these fossils date from the Anisian/Ladinian transition (about 237 million years ago).

Triassic–Jurassic extinction event

The Triassic period ended with a mass extinction, which was particularly severe in the oceans; the conodonts disappeared, as did all the marine reptiles except ichthyosaurs and plesiosaurs. Invertebrates like brachiopods, gastropods, and molluscs were severely affected. In the oceans, 22% of marine families and possibly about half of marine genera went missing.

Though the end-Triassic extinction event was not equally devastating in all terrestrial ecosystems, several important clades of crurotarsans (large archosaurian reptiles previously grouped together as the thecodonts) disappeared, as did most of the large labyrinthodont amphibians, groups of small reptiles, and some synapsids (except for the proto-mammals). Some of the early, primitive dinosaurs also became extinct, but more adaptive ones survived to evolve into the Jurassic. Surviving plants that went on to dominate the Mesozoic world included modern conifers and cycadeoids.

The cause of the Late Triassic extinction is uncertain. It was accompanied by huge volcanic eruptions that occurred as the supercontinent Pangaea began to break apart about 202 to 191 million years ago (40Ar/39Ar dates),[289] forming the Central Atlantic Magmatic Province (CAMP),[290] one of the largest known inland volcanic events since the planet had first cooled and stabilized. Other possible but less likely causes for the extinction events include global cooling or even a bolide impact, for which an impact crater containing Manicouagan Reservoir in Quebec, Canada, has been singled out. However, the Manicouagan impact melt has been dated to 214±1 Mya. The date of the Triassic-Jurassic boundary has also been more accurately fixed recently, at 201.3 Mya. Both dates are gaining accuracy by using more accurate forms of radiometric dating, in particular the decay of uranium to lead in zircons formed at time of the impact. So, the evidence suggests the Manicouagan impact preceded the end of the Triassic by approximately 10±2 Ma. It could, therefore, not be the immediate cause of the observed mass extinction.[291]

The number of Late Triassic extinctions is disputed. Some studies suggest that there are at least two periods of extinction towards the end of the Triassic, separated by 12 to 17 million years. But arguing against this is a recent study of North American faunas. In the Petrified Forest of northeast Arizona there is a unique sequence of late Carnian-early Norian terrestrial sediments. An analysis in 2002[292] found no significant change in the paleoenvironment. Phytosaurs, the most common fossils there, experienced a change-over only at the genus level, and the number of species remained the same. Some aetosaurs, the next most common tetrapods, and early dinosaurs, passed through unchanged. However, both phytosaurs and aetosaurs were among the groups of archosaur reptiles completely wiped out by the end-Triassic extinction event.

It seems likely then that there was some sort of end-Carnian extinction, when several herbivorous archosauromorph groups died out, while the large herbivorous therapsids— the kannemeyeriid dicynodonts and the traversodont cynodonts—were much reduced in the northern half of Pangaea (Laurasia).

These extinctions within the Triassic and at its end allowed the dinosaurs to expand into many niches that had become unoccupied. Dinosaurs became increasingly dominant, abundant and diverse, and remained that way for the next 150 million years. The true "Age of Dinosaurs" is during the following Jurassic and Cretaceous periods, rather than the Triassic.

References

- Emiliani, Cesare. (1992). *Planet Earth: Cosmology, Geology, & the Evolution of Life & the Environment*. Cambridge University Press. (Paperback Edition ISBN 0-521-40949-7)
- Ogg, Jim; June, 2004, *Overview of Global Boundary Stratotype Sections and Points (GSSP's)* Stratigraphy.org[293], Accessed April 30, 2006
- Stanley, Steven M. *Earth System History*. New York: W.H. Freeman and Company, 1999. ISBN 0-7167-2882-6
- Sues, Hans-Dieter & Fraser, Nicholas C. *Triassic Life on Land: The Great Transition* New York: Columbia University Press, 2010. Series: Critical Moments and Perspectives in Earth History and Paleobiology. ISBN 978-0-231-13522-1
- van Andel, Tjeerd, (1985) 1994, *New Views on an Old Planet: A History of Global Change*, Cambridge University Press

External links

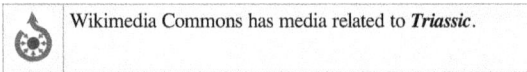

Wikimedia Commons has media related to *Triassic*.

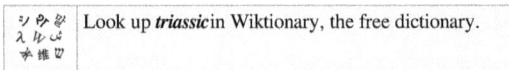

Look up *triassic* in Wiktionary, the free dictionary.

- A Visit To Ammonite Canyon, Nevada[294] Virtual field trip to one of the most famous places on Earth to study the Late Triassic extinction in a marine geologic section.
- Middle Triassic Ammonoids From Nevada[295]
- Ammonoids At Union Wash, California[296] Virtual field trip to a classic early Triassic ammonoid locality in Inyo County, California.
- Late Triassic Ichthyosaur And Invertebrate Fossils In Nevada[297]
- Early Triassic Ammonoid Fossils In Nevada[298]
- Overall introduction[299]
- 'The Triassic world'[300]
- Douglas Henderson's illustrations of Triassic animals[301] Wikipedia:Link rot
- Paleofiles page on the Triassic extinctions[302]
- Examples of Triassic Fossils[303]

Jurassic

Jurassic Period *201.3–145 million years ago* PreЄЄ OSD C P T J K PₘN	
Mean atmospheric O $_2$ content over period duration	c. 26 vol %[304,305] (130 % of modern level
Mean atmospheric CO $_2$ content over period duration	c. 1950 ppm[306] (7 times pre-industrial level)
Mean surface tempera- ture over period duration	c. 16.5 °C[307] (3 °C above modern level)

Key events in the Jurassic

An approximate timescale of key Jurassic events.
Vertical axis: millions of years ago.

The **Jurassic** (/dʒʊəˈræsɪk/; from Jura Mountains) was a geologic period and system that spanned 56 million years from the end of the Triassic Period 201.3 million years ago (Mya) to the beginning of the Cretaceous Period 145 Mya.[308] Víctor Ramos, one of the authors of the study proposing the 140 Ma boundary age, sees the study as a "first step" toward formally changing the age in the International Union of Geological Sciences.</ref> The Jurassic constitutes the middle period of the Mesozoic Era, also known as the Age of Reptiles.

The start of the period was marked by the major Triassic–Jurassic extinction event. Two other extinction events occurred during the period: the Pliens-bachian/Toarcian event in the Early Jurassic, and the Tithonian event at the end; however, neither event ranks among the "Big Five" mass extinctions.

The Jurassic period is divided into three epochs: Early, Middle, and Late. Similarly, in stratigraphy, the Jurassic is divided into the Lower Jurassic, Middle Jurassic, and Upper Jurassic series of rock formations.

The Jurassic is named after the Jura Mountains within the European Alps, where limestone strata from the period were first identified. By the beginning of the Jurassic, the supercontinent Pangaea had begun rifting into two land-masses: Laurasia to the north, and Gondwana to the south. This created more coastlines and shifted the continental climate from dry to humid, and many of the arid deserts of the Triassic were replaced by lush rainforests.

On land, the fauna transitioned from the Triassic fauna, dominated by both dinosauromorph and crocodylomorph archosaurs, to one dominated by di-nosaurs alone. The first birds also appeared during the Jurassic, having evolved from a branch of theropod dinosaurs. Other major events include the appear-ance of the earliest lizards, and the evolution of therian mammals, including primitive placentals. Crocodilians made the transition from a terrestrial to an aquatic mode of life. The oceans were inhabited by marine reptiles such as ichthyosaurs and plesiosaurs, while pterosaurs were the dominant flying verte-brates.

History of term

The chronostratigraphic term "Jurassic" is directly linked to the Jura Moun-tains, a mountain range mainly following the course of the France–Switzerland border. During a tour of the region in 1795,[309] Alexander von Humboldt rec-ognized the mainly limestone dominated mountain range of the Jura Moun-tains as a separate formation that had not been included in the established stratigraphic system defined by Abraham Gottlob Werner, and he named it "Jura-Kalkstein" ('Jura limestone') in 1799.[310,311,312,313]

Etymology

The name "Jura" is derived from the Celtic root *jor* via Gaulish *iuris* "wooded mountain", which, borrowed into Latin as a place name, evolved into *Juria* and finally *Jura*.[314]

Divisions

The Jurassic period is divided into three epochs: Early, Middle, and Late. Similarly, in stratigraphy, the Jurassic is divided into the Lower Jurassic, Middle Jurassic, and Upper Jurassic series of rock formations, also known as *Lias*, *Dogger* and *Malm* in Europe. The separation of the term **Jurassic** into three sections originated with Leopold von Buch. The faunal stages from youngest to oldest are:

Upper/Late Jurassic

Tithonian	$(152.1 \pm 4 - 145 \pm 4$ Mya)
Kimmeridgian	$(157.3 \pm 4 - 152.1 \pm 4$ Mya)
Oxfordian	$(163.5 \pm 4 - 157.3 \pm 4$ Mya)

Middle Jurassic

Callovian	$(166.1 \pm 4 - 163.5 \pm 4$ Mya)
Bathonian	$(168.3 \pm 3.5 - 166.1 \pm 4$ Mya)
Bajocian	$(170.3 \pm 3 - 168.3 \pm 3.5$ Mya)
Aalenian	$(174.1 \pm 2 - 170.3 \pm 3$ Mya)

Lower/Early Jurassic

Toarcian	$(182.7 \pm 1.5 - 174.1 \pm 2$ Mya)
Pliensbachian	$(190.8 \pm 1.5 - 182.7 \pm 1.5$ Mya)
Sinemurian	$(199.3 \pm 1 - 190.8 \pm 1.5$ Mya)
Hettangian	$(201.3 \pm 0.6 - 199.3 \pm 1$ Mya)

Paleogeography and tectonics

During the early Jurassic period, the supercontinent Pangaea broke up into the northern supercontinent Laurasia and the southern supercontinent Gondwana; the Gulf of Mexico opened in the new rift between North America and what is now Mexico's Yucatan Peninsula. The Jurassic North Atlantic Ocean was relatively narrow, while the South Atlantic did not open until the following Cretaceous period, when Gondwana itself rifted apart.[315] The Tethys Sea closed, and the Neotethys basin appeared. Climates were warm, with no evidence of glaciation. As in the Triassic, there was apparently no land over either pole, and no extensive ice caps existed.

The Jurassic geological record is good in western Europe, where extensive marine sequences indicate a time when much of that future landmass was submerged under shallow tropical seas; famous locales include the Jurassic

Figure 133: *Depiction of Early Jurassic environment pre-*
served at the St. George Dinosaur Discovery Site at Johnson
Farm, with Dilophosaurus wetherilli in bird-like resting pose

Coast World Heritage Site in southern England and the renowned late Jurassic *lagerstätten* of Holzmaden and Solnhofen in Germany. In contrast, the North American Jurassic record is the poorest of the Mesozoic, with few outcrops at the surface. Though the epicontinental Sundance Sea left marine deposits in parts of the northern plains of the United States and Canada during the late Jurassic, most exposed sediments from this period are continental, such as the alluvial deposits of the Morrison Formation.

The Jurassic was a time of calcite sea geochemistry in which low-magnesium calcite was the primary inorganic marine precipitate of calcium carbonate. Carbonate hardgrounds were thus very common, along with calcitic ooids, calcitic cements, and invertebrate faunas with dominantly calcitic skeletons (Stanley and Hardie, 1998, 1999).

The first of several massive batholiths were emplaced in the northern American cordillera beginning in the mid-Jurassic, marking the Nevadan orogeny.[316] Important Jurassic exposures are also found in Russia, India, South America, Japan, Australasia and the United Kingdom.

In Africa, Early Jurassic strata are distributed in a similar fashion to Late Triassic beds, with more common outcrops in the south and less common fossil beds which are predominated by tracks to the north. As the Jurassic proceeded, larger and more iconic groups of dinosaurs like sauropods and ornithopods proliferated in Africa. Middle Jurassic strata are neither well represented nor well studied in Africa. Late Jurassic strata are also poorly represented apart from the spectacular Tendaguru fauna in Tanzania. The Late Jurassic life of Tendaguru is very similar to that found in western North America's Morrison Formation.[317]

Figure 134: *Jurassic limestones and marls (the Matmor Formation) in southern Israel.*

Figure 135: *The late Jurassic Morrison Formation in Colorado is one of the most fertile sources of dinosaur fossils in North America.*

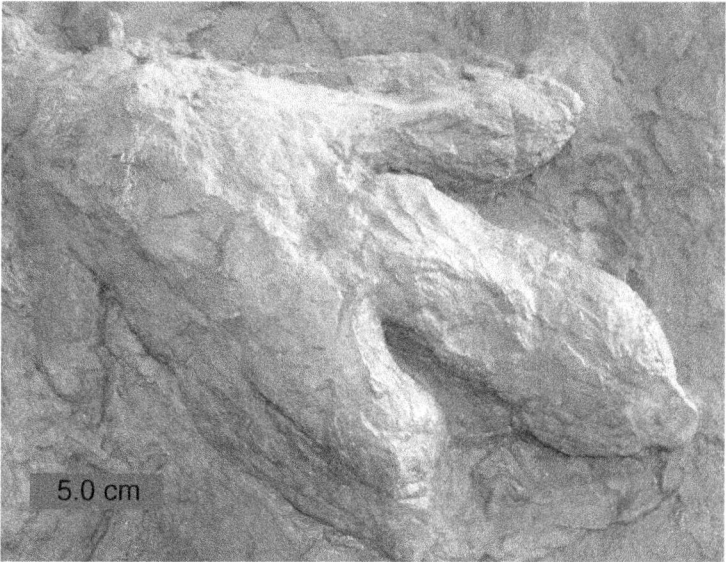

Figure 136: *Gigandipus, a dinosaur footprint in the Lower Jurassic Moenave Formation at the St. George Dinosaur Discovery Site at Johnson Farm, southwestern Utah.*

Figure 137: *The Permian through Jurassic stratigraphy of the Colorado Plateau area of southeastern Utah.*

Fauna

Aquatic and marine

During the Jurassic period, the primary vertebrates living in the sea were fish and marine reptiles. The latter include ichthyosaurs, which were at the peak of their diversity, plesiosaurs, pliosaurs, and marine crocodiles of the families Teleosauridae and Metriorhynchidae.[318] Numerous turtles could be found in lakes and rivers.

In the invertebrate world, several new groups appeared, including rudists (a reef-forming variety of bivalves) and belemnites. Calcareous sabellids (*Glomerula*) appeared in the Early Jurassic. The Jurassic also had diverse encrusting and boring (sclerobiont) communities, and it saw a significant rise in the bioerosion of carbonate shells and hardgrounds. Especially common is the ichnogenus (trace fossil) *Gastrochaenolites*.

During the Jurassic period, about four or five of the twelve clades of plank-tonic organisms that exist in the fossil record either experienced a massive evolutionary radiation or appeared for the first time.

Figure 138: *A Pliosaurus (right) harassing a Leedsichthys in a Jurassic sea.*

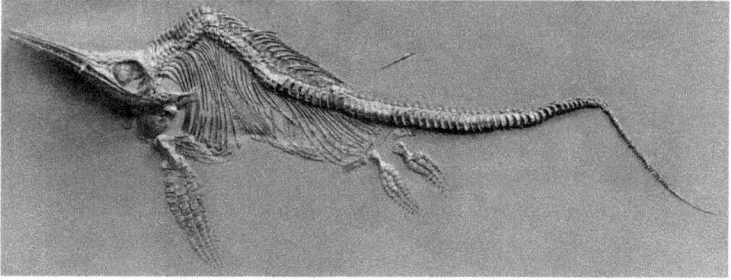

Figure 139: *Ichthyosaurus from lower (early) Jurassic slates in southern Germany featured a dolphin-like body shape.*

Figure 140: *Plesiosaurs like Muraenosaurus roamed Jurassic oceans.*

Figure 141: *Gastropod and attached mytilid bivalves on a Jurassic limestone bedding plane in southern Israel.*

Figure 142: *Example of Rare Early Jurassic (Pliensbachian) Ecosystem,
the Drzewica Formation of Szydłowiec, Poland. This zone was charac-
terised by a very damp ecosystem populated by dinosaur megafauna,
more related in several aspects to Middle or Late Jurassic dinosaurs*

Terrestrial

On land, various archosaurian reptiles remained dominant. The Jurassic
was a golden age for the large herbivorous dinosaurs known as the sauropods—*Camarasaurus, Apatosaurus, Diplodocus, Brachiosaurus*, and many
others—that roamed the land late in the period; their foraging grounds were
either the prairies of ferns, palm-like cycads and bennettitales, or the higher
coniferous growth, according to their adaptations. The smaller Ornithischian
herbivore dinosaurs, like stegosaurs and small ornithopods were less predominant, but played important roles. They were preyed upon by large theropods,
such as *Ceratosaurus, Megalosaurus, Torvosaurus* and *Allosaurus*. All these
belong to the 'lizard hipped' or saurischian branch of the dinosaurs. During
the Late Jurassic, the first avialans, like *Archaeopteryx*, evolved from small
coelurosaurian dinosaurs. In the air, pterosaurs were common; they ruled the
skies, filling many ecological roles now taken by birds, and may have already
produced some of the largest flying animals of all time.[319] Within the undergrowth were various types of early mammals, as well as tritylodonts, lizard-like
sphenodonts, and early lissamphibians. The rest of the Lissamphibia evolved
in this period, introducing the first salamanders and caecilians.

Figure 143: *Diplodocus, reaching lengths over 30 m, was a common sauropod during the late Jurassic.*

Figure 144: *Allosaurus was one of the largest land predators during the Jurassic.*

Figure 145: *Stegosaurus is one of the most recognizable genera of dinosaurs and lived during the mid to late Jurassic.*

Figure 146: *Archaeopteryx, a primitive bird-like dinosaur, appeared in the Late Jurassic.*

Figure 147: *Aurornis xui, which lived in the late Jurassic, may be the most primitive avialan dinosaur known to date, and is one of the earliest avialans found to date.*

Flora

The arid, continental conditions characteristic of the Triassic steadily eased during the Jurassic period, especially at higher latitudes; the warm, humid climate allowed lush jungles to cover much of the landscape.[320] Gymnosperms were relatively diverse during the Jurassic period. The Conifers in particular dominated the flora, as during the Triassic; they were the most diverse group and constituted the majority of large trees.

Extant conifer families that flourished during the Jurassic included the Araucariaceae, Cephalotaxaceae, Pinaceae, Podocarpaceae, Taxaceae and Taxodiaceae.[321] The extinct Mesozoic conifer family Cheirolepidiaceae dominated low latitude vegetation, as did the shrubby Bennettitales.[322] Cycads, similar to palm trees, were also common, as were ginkgos and Dicksoniaceous tree ferns in the forest. Smaller ferns were probably the dominant undergrowth. Caytoniaceous seed ferns were another group of important plants during this time and are thought to have been shrub to small-tree sized.[323] Ginkgo plants were particularly common in the mid- to high northern latitudes. In the Southern Hemisphere, podocarps were especially successful, while Ginkgos and Czekanowskiales were rare.

In the oceans, modern coralline algae appeared for the first time. However, they were a part of another major extinction that happened within the next major time period.

Figure 148: *Conifers were the dominant land plants of the Jurassic*

Figure 149: *Various dinosaurs roamed forests of similarly large conifers during the Jurassic period.*

References

- Behrensmeyer, Damuth, J.D., DiMichele, W.A., Potts, R., Sues, H.D. & Wing, S.L. (eds.) (1992), *Terrestrial Ecosystems through Time: the Evolutionary Paleoecology of Terrestrial Plants and Animals*, University of Chicago Press, Chicago and London, ISBN 0-226-04154-9 (cloth), ISBN 0-226-04155-7 (paper).
- Haines, Tim (2000) *Walking with Dinosaurs: A Natural History*, New York: Dorling Kindersley Publishing, Inc., p. 65. ISBN 0-563-38449-2.
- Kazlev, M. Alan (2002) Palaeos website[324] Accessed Jan. 8, 2006.
- Mader, Sylvia (2004) Biology, eighth edition.
- Monroe, James S., and Reed Wicander. (1997) *The Changing Earth: Exploring Geology and Evolution*, 2nd ed. Belmont: West Publishing Company, 1997. ISBN 0-314-09577-2.
- Ogg, Jim; June, 2004, *Overview of Global Boundary Stratotype Sections and Points (GSSP's)*, International Commission on Stratigraphy, pp. 17
- Stanley, S.M.; Hardie, L.A. (1998). "Secular oscillations in the carbonate mineralogy of reef-building and sediment-producing organisms driven by tectonically forced shifts in seawater chemistry". *Palaeogeography, Palaeoclimatology, Palaeoecology*. 144: 3–19. Bibcode: 1998PPP...144....3S[325]. doi: 10.1016/s0031-0182(98)00109-6[326].
- Stanley, S.M.; Hardie, L.A. (1999). "Hypercalcification; paleontology links plate tectonics and geochemistry to sedimentology". *GSA Today*. 9: 1–7.
- Taylor, P.D.; Wilson, M.A. (2003). "Palaeoecology and evolution of marine hard substrate communities"[327] (PDF). *Earth-Science Reviews*. 62: 1–103. Bibcode: 2003ESRv...62....1T[328]. doi: 10.1016/s0012-8252(02)00131-9[329]. Archived from the original[330] (PDF) on 2009-03-25.

External links

Wikimedia Commons has media related to *Jurassic*.

シ ゆ 釟	Look up *jurassic* in Wiktionary, the free dictionary.
λ ル ぶ	
氺 推 ヷ	

- Examples of Jurassic Fossils[331]
- Jurassic fossils in Harbury, Warwickshire[332]
- Jurassic Microfossils: 65+ images of Foraminifera[333]

<indicator name="spoken-icon"> ◉)) </indicator>

Cretaceous

Cretaceous Period *145–66 million years ago* PreЄЄ OSD C P T J K PgN	
Mean atmospheric O$_2$ content over period duration	c. 30 vol %[334,335] (150 % of modern level
Mean atmospheric CO$_2$ content over period duration	c. 1700 ppm[336] (6 times pre-industrial level)
Mean surface temperature over period duration	c. 18 °C[337] (4 °C above modern level)

Key events in the Cretaceous

```
  70
  80
  90
 100
 110
 120
 130
 140
```

An approximate timescale of key Cretaceous events.
Axis scale: millions of years ago.

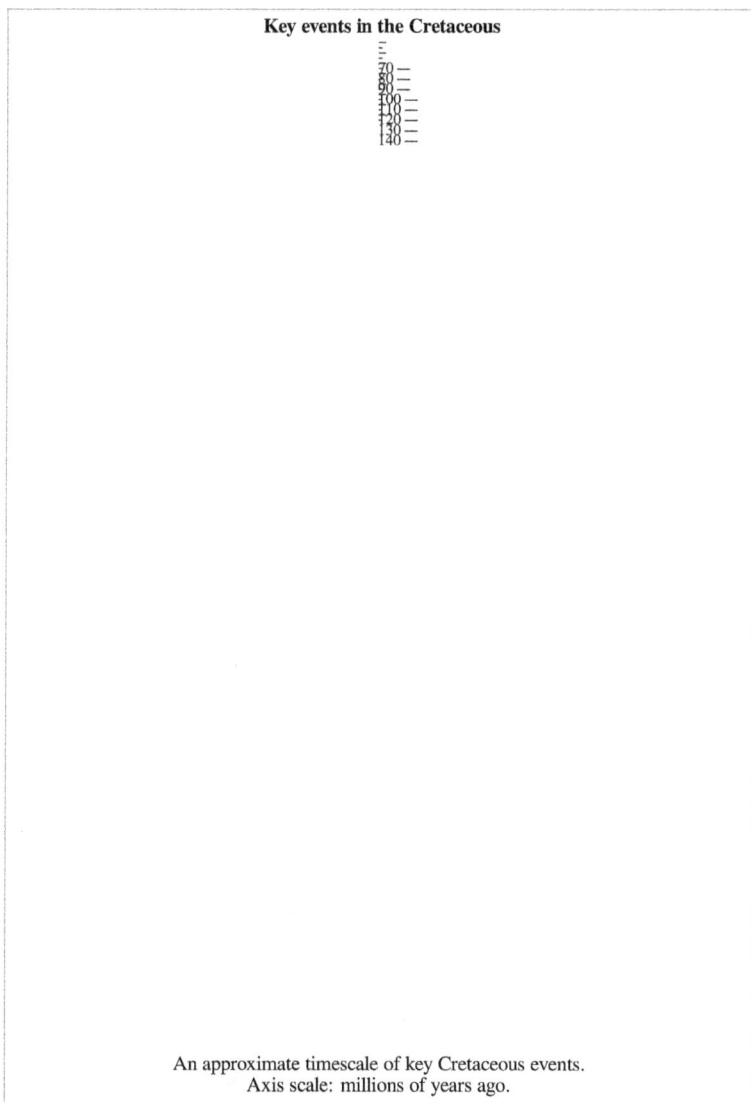

The **Cretaceous** (/krɪˈteɪʃəs/, *kri-TAY-shəs*) is a geologic period and system that spans 79 million years from the end of the Jurassic Period 145 million years ago (mya) to the beginning of the Paleogene Period 66 mya. It is the last period of the Mesozoic Era, and the longest period of the Phanerozoic Eon. The Cretaceous Period is usually abbreviated **K**, for its German translation *Kreide* (chalk).

The Cretaceous was a period with a relatively warm climate, resulting in high eustatic sea levels that created numerous shallow inland seas. These oceans and seas were populated with now-extinct marine reptiles, ammonites and rudists, while dinosaurs continued to dominate on land. During this time, new groups of mammals and birds, as well as flowering plants, appeared.

The Cretaceous (along with the Mesozoic) ended with the Cretaceous–Paleogene extinction event, a large mass extinction in which many groups, including non-avian dinosaurs, pterosaurs and large marine reptiles died out. The end of the Cretaceous is defined by the abrupt Cretaceous–Paleogene boundary (K–Pg boundary), a geologic signature associated with the mass extinction which lies between the Mesozoic and Cenozoic eras.

Geology

Research history

The Cretaceous as a separate period was first defined by Belgian geologist Jean d'Omalius d'Halloy in 1822,[338] using strata in the Paris Basin and named for the extensive beds of chalk (calcium carbonate deposited by the shells of marine invertebrates, principally coccoliths), found in the upper Cretaceous of Western Europe.[339] The name Cretaceous was derived from Latin *creta*, meaning *chalk*.

Stratigraphic subdivisions

The Cretaceous is divided into Early and Late Cretaceous epochs, or Lower and Upper Cretaceous series. In older literature the Cretaceous is sometimes divided into three series: Neocomian (lower/early), Gallic (middle) and Senonian (upper/late). A subdivision in eleven stages, all originating from European stratigraphy, is now used worldwide. In many parts of the world, alternative local subdivisions are still in use.

As with other older geologic periods, the rock beds of the Cretaceous are well identified but the exact age of the system's base is uncertain by a few million years. No great extinction or burst of diversity separates the Cretaceous from the Jurassic. However, the top of the system is sharply defined, being placed at an iridium-rich layer found worldwide that is believed to be associated with the Chicxulub impact crater, with its boundaries circumscribing parts of the Yucatán Peninsula and into the Gulf of Mexico. This layer has been dated at 66.043 Ma.

A 140 Ma age for the Jurassic-Cretaceous boundary instead of the usually accepted 145 Ma was proposed in 2014 based on a stratigraphic study of Vaca Muerta Formation in Neuquén Basin, Argentina. Víctor Ramos, one of the authors of the study proposing the 140 Ma boundary age sees the study as a "first step" toward formally changing the age in the International Union of Geological Sciences.

From youngest to oldest, the subdivisions of the Cretaceous period are:

Late Cretaceous

Maastrichtian – (66-72.1 MYA)

Campanian – (72.1-83.6 MYA)

Santonian – (83.6-86.3 MYA)

Coniacian – (86.3-89.8 MYA)

Turonian – (89.8-93.9 MYA)

Cenomanian – (93.9-100.5 MYA)

Early Cretaceous

Albian – (100.5-113.0 MYA)

Aptian – (113.0-125.0 MYA)

Barremian – (125.0-129.4 MYA)

Hauterivian – (129.4-132.9 MYA)

Valanginian – (132.9-139.8 MYA)

Berriasian – (139.8-145.0 MYA)

Rock formations

The high sea level and warm climate of the Cretaceous meant large areas of the continents were covered by warm, shallow seas, providing habitat for many marine organisms. The Cretaceous was named for the extensive chalk deposits of this age in Europe, but in many parts of the world, the deposits from the Cretaceous are of marine limestone, a rock type that is formed under warm, shallow marine circumstances. Due to the high sea level, there was extensive space for such sedimentation. Because of the relatively young age and great thickness of the system, Cretaceous rocks are evident in many areas worldwide.

Chalk is a rock type characteristic for (but not restricted to) the Cretaceous. It consists of coccoliths, microscopically small calcite skeletons of coccolithophores, a type of algae that prospered in the Cretaceous seas.

In northwestern Europe, chalk deposits from the Upper Cretaceous are characteristic for the Chalk Group, which forms the white cliffs of Dover on the south coast of England and similar cliffs on the French Normandian coast. The

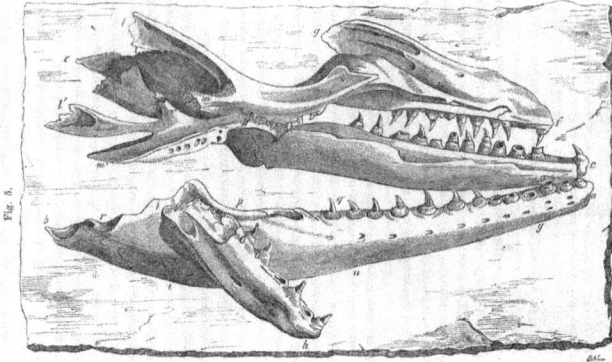

Figure 150: *Drawing of fossil jaws of Mosasaurus hoffmanni, from the Maastrichtian of Dutch Limburg, by Dutch geologist Pieter Harting (1866).*

Figure 151: *Scipionyxs complete slab*

group is found in England, northern France, the low countries, northern Germany, Denmark and in the subsurface of the southern part of the North Sea. Chalk is not easily consolidated and the Chalk Group still consists of loose sediments in many places. The group also has other limestones and arenites. Among the fossils it contains are sea urchins, belemnites, ammonites and sea reptiles such as *Mosasaurus*.

In southern Europe, the Cretaceous is usually a marine system consisting of competent limestone beds or incompetent marls. Because the Alpine mountain chains did not yet exist in the Cretaceous, these deposits formed on the southern edge of the European continental shelf, at the margin of the Tethys Ocean.

Stagnation of deep sea currents in middle Cretaceous times caused anoxic conditions in the sea water leaving the deposited organic matter undecomposed. Half the worlds petroleum reserves were laid down at this time in the anoxic conditions of what would become the Persian Gulf and the Gulf of Mexico. In many places around the world, dark anoxic shales were formed during this interval.[340] These shales are an important source rock for oil and gas, for example in the subsurface of the North Sea.

Paleogeography

During the Cretaceous, the late-Paleozoic-to-early-Mesozoic supercontinent of Pangaea completed its tectonic breakup into the present-day continents, although their positions were substantially different at the time. As the Atlantic Ocean widened, the convergent-margin mountain building (orogenies) that had begun during the Jurassic continued in the North American Cordillera, as the Nevadan orogeny was followed by the Sevier and Laramide orogenies.

Though Gondwana was still intact in the beginning of the Cretaceous, it broke up as South America, Antarctica and Australia rifted away from Africa (though India and Madagascar remained attached to each other); thus, the South Atlantic and Indian Oceans were newly formed. Such active rifting lifted great undersea mountain chains along the welts, raising eustatic sea levels worldwide. To the north of Africa the Tethys Sea continued to narrow. Broad shallow seas advanced across central North America (the Western Interior Seaway) and Europe, then receded late in the period, leaving thick marine deposits sandwiched between coal beds. At the peak of the Cretaceous transgression, one-third of Earth's present land area was submerged.[341]

The Cretaceous is justly famous for its chalk; indeed, more chalk formed in the Cretaceous than in any other period in the Phanerozoic.[342] Mid-ocean ridge activity—or rather, the circulation of seawater through the enlarged

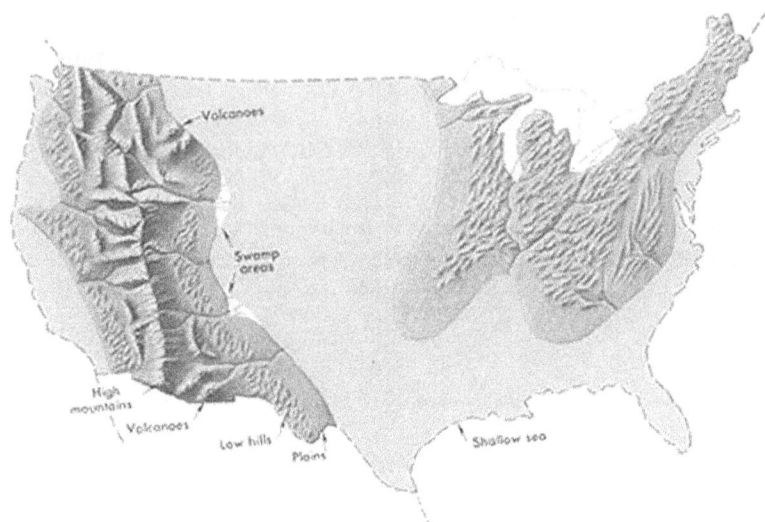

Figure 152: *Geography of the Contiguous United States in the late Cretaceous period*

ridges—enriched the oceans in calcium; this made the oceans more saturated, as well as increased the bioavailability of the element for calcareous nanoplankton.[343] These widespread carbonates and other sedimentary deposits make the Cretaceous rock record especially fine. Famous formations from North America include the rich marine fossils of Kansas's Smoky Hill Chalk Member and the terrestrial fauna of the late Cretaceous Hell Creek Formation. Other important Cretaceous exposures occur in Europe (e.g., the Weald) and China (the Yixian Formation). In the area that is now India, massive lava beds called the Deccan Traps were erupted in the very late Cretaceous and early Paleocene.

Climate

The cooling trend of last epoch of the Jurassic continued into the first age of the Cretaceous. There is evidence that snowfalls were common in the higher latitudes and the tropics became wetter than during the Triassic and Jurassic. Glaciation was however restricted to high-latitude mountains, though seasonal snow may have existed farther from the poles. Rafting by ice of stones into marine environments occurred during much of the Cretaceous but evidence of deposition directly from glaciers is limited to the Early Cretaceous of the Eromanga Basin in southern Australia.

After the end of the first age, however, temperatures increased again, and these conditions were almost constant until the end of the period. The warming may have been due to intense volcanic activity which produced large quantities of carbon dioxide. Between 70–69 Ma and 66–65 Ma, isotopic ratios indicate elevated atmospheric CO_2 pressures with levels of 1000–1400 ppmV and mean annual temperatures in west Texas between 21 and 23 °C (70-73 °F). Atmospheric CO_2 and temperature relations indicate a doubling of pCO_2 was accompanied by a ~0.6 °C increase in temperature.[344] The production of large quantities of magma, variously attributed to mantle plumes or to extensional tectonics, further pushed sea levels up, so that large areas of the continental crust were covered with shallow seas. The Tethys Sea connecting the tropical oceans east to west also helped to warm the global climate. Warm-adapted plant fossils are known from localities as far north as Alaska and Greenland, while dinosaur fossils have been found within 15 degrees of the Cretaceous south pole.[345]

Nonetheless, there is evidence of Antarctic marine glaciation in the Turonian Age.

A very gentle temperature gradient from the equator to the poles meant weaker global winds, which drive the ocean currents, resulted in less upwelling and more stagnant oceans than today. This is evidenced by widespread black shale deposition and frequent anoxic events.[346] Sediment cores show that tropical sea surface temperatures may have briefly been as warm as 42 °C (108 °F), 17 °C (31 °F) warmer than at present, and that they averaged around 37 °C (99 °F). Meanwhile, deep ocean temperatures were as much as 15 to 20 °C (27 to 36 °F) warmer than today's.[347,348]

Life

Flora

Flowering plants (angiosperms) spread during this period, although they did not become predominant until the Campanian Age near the end of the period. Their evolution was aided by the appearance of bees; in fact angiosperms and insects are a good example of coevolution. The first representatives of many leafy trees, including figs, planes and magnolias, appeared in the Cretaceous. At the same time, some earlier Mesozoic gymnosperms continued to thrive; pehuéns (monkey puzzle trees, *Araucaria*) and other conifers being notably plentiful and widespread. Some fern orders such as Gleicheniales[349] appeared as early in the fossil record as the Cretaceous and achieved an early broad distribution. Gymnosperm taxa like Bennettitales and hirmerellan conifers died out before the end of the period.

Figure 153: *Although the first representatives of leafy trees and true grasses emerged in the Cretaceous, the flora was still dominated by conifers like Araucaria (Here: Modern Araucaria araucana in Chile).*

Terrestrial fauna

On land, mammals were generally small sized, but a very relevant component of the fauna, with cimolodont multituberculates outnumbering dinosaurs in some sites.[350] Neither true marsupials nor placentals existed until the very end, but a variety of non-marsupial metatherians and non-placental eutherians had already begun to diversify greatly, ranging as carnivores (Deltatheroida), aquatic foragers (Stagodontidae) and herbivores (*Schowalteria*, Zhelestidae). Various "archaic" groups like eutriconodonts were common in the Early Cretaceous, but by the Late Cretaceous northern mammalian faunas were dominated by multituberculates and therians, with dryolestoids dominating South America.

The apex predators were archosaurian reptiles, especially dinosaurs, which were at their most diverse stage. Pterosaurs were common in the early and middle Cretaceous, but as the Cretaceous proceeded they declined for poorly understood reasons (once thought to be due to competition with early birds, but now it is understood avian adaptive radiation is not consistent with pterosaur decline), and by the end of the period only two highly specialized families remained.

The Liaoning lagerstätte (Chaomidianzi formation) in China is a treasure chest of preserved remains of numerous types of small dinosaurs, birds and mammals, that provides a glimpse of life in the Early Cretaceous. The coelurosaur dinosaurs found there represent types of the group Maniraptora, which is transitional between dinosaurs and birds, and are notable for the presence of hair-like feathers.

Insects diversified during the Cretaceous, and the oldest known ants, termites and some lepidopterans, akin to butterflies and moths, appeared. Aphids, grasshoppers and gall wasps appeared.

Figure 154: *Tyrannosaurus rex, one of the largest land predators of all time, lived during the late Cretaceous.*

Figure 155: *Up to 2 m long and 0.5 m high at the hip, Velociraptor was feathered and roamed the late Cretaceous.*

Figure 156: *Triceratops, one of the most recognizable genera of the Cretaceous*

Figure 157: *A pterosaur, Anhanguera piscator*

Figure 158: *Confuciusornis, a genus of crow-sized birds from the Early Cretaceous*

Figure 159: *Ichthyornis was a toothed seabird-
like ornithuran from the late Cretaceous period*

Marine fauna

In the seas, rays, modern sharks and teleosts became common. Marine reptiles included ichthyosaurs in the early and mid-Cretaceous (becoming extinct
during the late Cretaceous Cenomanian-Turonian anoxic event), plesiosaurs
throughout the entire period, and mosasaurs appearing in the Late Cretaceous.

Baculites, an ammonite genus with a straight shell, flourished in the seas along
with reef-building rudist clams. The Hesperornithiformes were flightless, marine diving birds that swam like grebes. Globotruncanid Foraminifera and
echinoderms such as sea urchins and starfish (sea stars) thrived. The first radiation of the diatoms (generally siliceous shelled, rather than calcareous) in the
oceans occurred during the Cretaceous; freshwater diatoms did not appear until the Miocene. The Cretaceous was also an important interval in the evolution
of bioerosion, the production of borings and scrapings in rocks, hardgrounds
and shells.

Figure 160: *A scene from the early Cretaceous: a Woolungasaurus is attacked by a Kronosaurus.*

Figure 161: *Tylosaurus was a large mosasaur, carnivorous marine reptiles that emerged in the late Cretaceous.*

Figure 162: *Strong-swimming and toothed predatory waterbird Hesperornis roamed late Cretacean oceans.*

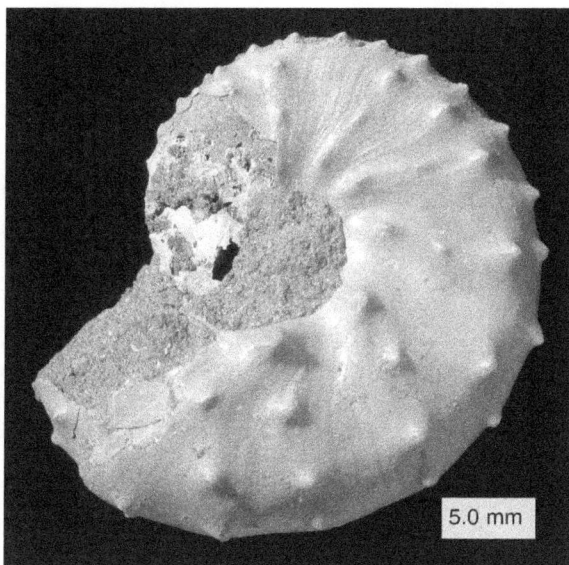

Figure 163: *The ammonite Discoscaphites iris, Owl Creek Formation (Upper Cretaceous), Ripley, Mississippi*

Figure 164: *A plate with Nematonotus sp., Pseudostacus sp. and a partial Dercetis triqueter, found in Hakel, Lebanon*

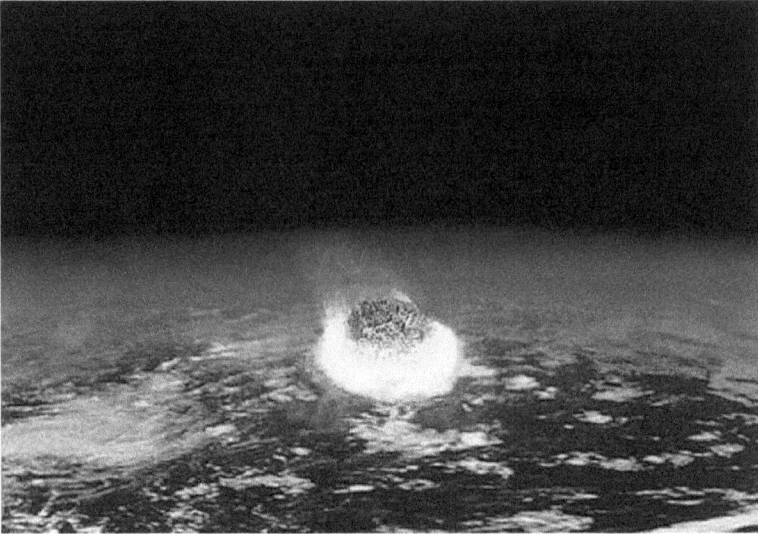

Figure 165: *The impact of a meteorite or comet is today widely accepted as the main reason for the Cretaceous–Paleogene extinction event.*

End-Cretaceous extinction event

The impact of a large body with the Earth may have been the punctuation mark at the end of a progressive decline in biodiversity during the Maastrichtian Age of the Cretaceous Period. The result was the extinction of three-quarters of Earth's plant and animal species. The impact created the sharp break known as K–Pg boundary (formerly known as the K–T boundary). Earth's biodiversity required substantial time to recover from this event, despite the probable existence of an abundance of vacant ecological niches.

Despite the severity of K-Pg extinction event, there was significant variability in the rate of extinction between and within different clades. Species which depended on photosynthesis declined or became extinct as atmospheric particles blocked solar energy. As is the case today, photosynthesizing organisms, such as phytoplankton and land plants, formed the primary part of the food chain in the late Cretaceous, and all else that depended on them suffered as well. Herbivorous animals, which depended on plants and plankton as their food, died out as their food sources became scarce; consequently, the top predators such as *Tyrannosaurus rex* also perished. Yet only three major groups of tetrapods disappeared completely; the non-avian dinosaurs, the plesiosaurs and the pterosaurs. The other Cretaceous groups that did not survive into the

Cenozoic era, the ichthyosaurs and last remaining temnospondyls and non-mammalian cynodonts were already extinct millions of years before the event occurred.Wikipedia:Citation needed

Coccolithophorids and molluscs, including ammonites, rudists, freshwater snails and mussels, as well as organisms whose food chain included these shell builders, became extinct or suffered heavy losses. For example, it is thought that ammonites were the principal food of mosasaurs, a group of giant marine reptiles that became extinct at the boundary.

Omnivores, insectivores and carrion-eaters survived the extinction event, perhaps because of the increased availability of their food sources. At the end of the Cretaceous there seem to have been no purely herbivorous or carnivorous mammals. Mammals and birds which survived the extinction fed on insects, larvae, worms and snails, which in turn fed on dead plant and animal matter. Scientists theorise that these organisms survived the collapse of plant-based food chains because they fed on detritus.

In stream communities, few groups of animals became extinct. Stream communities rely less on food from living plants and more on detritus that washes in from land. This particular ecological niche buffered them from extinction. Similar, but more complex patterns have been found in the oceans. Extinction was more severe among animals living in the water column, than among animals living on or in the seafloor. Animals in the water column are almost entirely dependent on primary production from living phytoplankton, while animals living on or in the ocean floor feed on detritus or can switch to detritus feeding.

The largest air-breathing survivors of the event, crocodilians and champsosaurs, were semi-aquatic and had access to detritus. Modern crocodilians can live as scavengers and can survive for months without food and go into hibernation when conditions are unfavorable, and their young are small, grow slowly, and feed largely on invertebrates and dead organisms or fragments of organisms for their first few years. These characteristics have been linked to crocodilian survival at the end of the Cretaceous.

Figure 166: *Numerous borings in a Cretaceous cobble, Faringdon, England; these are excellent examples of fossil bioerosion.*

Figure 167: *Cretaceous hardground from Texas with encrusting oysters and borings. The scale bar is 10 mm.*

Figure 168: *Rudist bivalves from the Cretaceous of the Omani Mountains, United Arab Emirates. Scale bar is 10 mm.*

Figure 169: *Inoceramus from the Cretaceous of South Dakota.*

Bibliography

- Yuichiro Kashiyama; Nanako O. Ogawa; Junichiro Kuroda; Motoo Shiro; Shinya Nomoto; Ryuji Tada; Hiroshi Kitazato; Naohiko Ohkouchi (May 2008). "Diazotrophic cyanobacteria as the major photoautotrophs during mid-Cretaceous oceanic anoxic events: Nitrogen and carbon isotopic evidence from sedimentary porphyrin"[351]. *Organic Geochemistry*. **39** (5): 532–549. doi: 10.1016/j.orggeochem.2007.11.010[352].
- Neal L Larson, Steven D Jorgensen, Robert A Farrar and Peter L Larson. *Ammonites and the other Cephalopods of the Pierre Seaway*. Geoscience Press, 1997.
- Ogg, Jim; June, 2004, *Overview of Global Boundary Stratotype Sections and Points (GSSP's)* https://web.archive.org/web/20060716071827/http://www.stratigraphy.org/gssp.htm Accessed April 30, 2006.
- Ovechkina, M.N.; Alekseev, A.S. (2005). "Quantitative changes of calcareous nannoflora in the Saratov region (Russian Platform) during the late Maastrichtian warming event"[353] (PDF). *Journal of Iberian Geology*. **31** (1): 149–165. Archived from the original[354] (PDF) on August 24, 2006.
- Rasnitsyn, A.P. and Quicke, D.L.J. (2002). *History of Insects*. Kluwer Academic Publishers. ISBN 1-4020-0026-X.—detailed coverage of various aspects of the evolutionary history of the insects.
- Skinner, Brian J., and Stephen C. Porter. *The Dynamic Earth: An Introduction to Physical Geology*. 3rd ed. New York: John Wiley & Sons, Inc., 1995. ISBN 0-471-60618-9
- Stanley, Steven M. *Earth System History*. New York: W.H. Freeman and Company, 1999. ISBN 0-7167-2882-6
- Taylor, P. D.; Wilson, M. A. (2003). "Palaeoecology and evolution of marine hard substrate communities". *Earth-Science Reviews*. **62**: 1–103. Bibcode: 2003ESRv...62....1T[355]. doi: 10.1016/S0012-8252(02)00131-9[356].

External links

Wikimedia Commons has media related to *Cretaceous*.

| ⸰⸰⸰ | Look up *cretaceous* in Wiktionary, the free dictionary. |

- UCMP Berkeley Cretaceous page[357]
- Bioerosion website at The College of Wooster[358]
- Cretaceous Microfossils: 180+ images of Foraminifera[359]
- ⊚ "Cretaceous System". *Encyclopædia Britannica*. **7** (11th ed.). 1911. pp. 414–418.

Cenozoic

| **Cenozoic Era** |
| *66 - 0 million years ago* |

Events of the Cenozoic

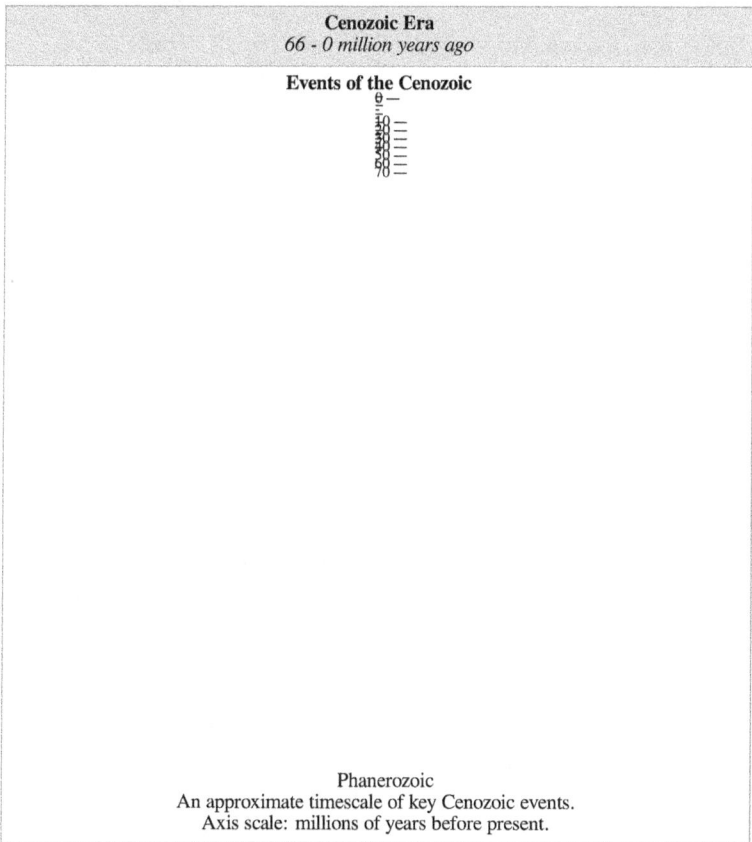

Phanerozoic
An approximate timescale of key Cenozoic events.
Axis scale: millions of years before present.

The **Cenozoic** Era (/ˌsiːnəˈzoʊɪk, <wbr />ˌsɛ-/)[360,361] meaning "new life", is the current and most recent of the three Phanerozoic geological eras, following the Mesozoic Era and, extending from 66 million years ago to the present day.

The Cenozoic is also known as the *Age of Mammals,* because the extinction of many groups allowed mammals to greatly diversify so that large mammals dominated it. The continents also moved into their current positions during this era.

Early in the Cenozoic, following the K-Pg extinction event, most of the fauna was relatively small, and included small mammals, birds, reptiles, and amphibians. From a geological perspective, it did not take long for mammals and birds to greatly diversify in the absence of the large reptiles that had dominated during the Mesozoic. A group of avians became known as the "terror birds," grew larger than the average human and were formidable predators. Mammals came to occupy almost every available niche (both marine and terrestrial), and some also grew very large, attaining sizes not seen in most of today's mammals.

Climate-wise, the Earth had begun a drying and cooling trend, culminating in the glaciations of the Pleistocene Epoch, and partially offset by the Paleocene-Eocene Thermal Maximum. The continents also began looking roughly familiar at this time and moved into their current positions.

Nomenclature

Cenozoic, meaning "new life," is derived from Greek καινός *kainós* "new," and ζωή *zōé* "life." The era is also known as the *Cænozoic, Caenozoic,* or *Cainozoic* (/ˌkaɪnəˈzoʊɪk, <wbr />ˌkeɪ-/).[362,363] The name "Cenozoic" (originally: "Kainozoic") was proposed in 1840 by the British geologist John Phillips (1800–1874).[364] Wikipedia:No original research#Primary, secondary and tertiary sources

Divisions

The Cenozoic is divided into three periods: the Paleogene, Neogene, and Quaternary; and seven epochs: the Paleocene, Eocene, Oligocene, Miocene, Pliocene, Pleistocene, and Holocene. The Quaternary Period was officially recognized by the International Commission on Stratigraphy in June 2009, and the former term, Tertiary Period, became officially disused in 2004 due to the need to divide the Cenozoic into periods more like those of the earlier Paleozoic and Mesozoic eras.[365] Wikipedia:Please clarify The common use of epochs during the Cenozoic helps paleontologists better organize and group the many significant events that occurred during this comparatively short interval of time. Knowledge of this era is relatively more detailed than any other era because of the relatively young, well-preserved rocks associated with it.

Figure 170: *Basilosaurus*

Paleogene Period

The Paleogene spans from the extinction of non-avian dinosaurs, 66 million years ago, to the dawn of the Neogene, 23.03 million years ago. It features three epochs: the Paleocene, Eocene and Oligocene.

The Paleocene epoch lasted from 66 million to 56 million years ago. Modern placental mammals originated during this time. The Paleocene is a transitional point between the devastation that is the K-T extinction, to the rich jungle environment that is the Early Eocene. The Early Paleocene saw the recovery of the earth. The continents began to take their modern shape, but all the continents and the subcontinent of India were separated from each other. Afro-Eurasia was separated by the Tethys Sea, and the Americas were separated by the strait of Panama, as the isthmus had not yet formed. This epoch featured a general warming trend, with jungles eventually reaching the poles. The oceans were dominated by sharks as the large reptiles that had once predominated were extinct. Archaic mammals filled the world such as creodonts (extinct carnivores, unrelated to existing Carnivora).

The Eocene Epoch ranged from 56 million years to 33.9 million years ago. In the Early-Eocene, species living in dense forest were unable to evolve into larger forms, as in the Paleocene. There was nothing over the weight of 10 kilograms. Among them were early primates, whales and horses along with many other early forms of mammals. At the top of the food chains were huge birds, such as Paracrax. The temperature was 30 degrees Celsius with little temperature gradient from pole to pole. In the Mid-Eocene, the Circumpolar-Antarctic current between Australia and Antarctica formed. This disrupted ocean currents worldwide and as a result caused a global cooling effect, shrinking the jungles. This allowed mammals to grow to mammoth proportions, such as whales which, by that time, had become almost fully aquatic. Mammals like *Andrewsarchus* were at the top of the food-chain. The Late Eocene saw the rebirth of seasons, which caused the expansion of savanna-like areas, along with the evolution of grass. The end of the Eocene was marked by the Eocene-Oligocene extinction event, the European face of which is known as the Grande Coupure.

The Oligocene Epoch spans from 33.9 million to 23.03 million years ago. The Oligocene featured the expansion of grass which had led to many new species to evolve, including the first elephants, cats, dogs, marsupials and many other species still prevalent today. Many other species of plants evolved in this period too. A cooling period featuring seasonal rains was still in effect. Mammals still continued to grow larger and larger.

Neogene

The Neogene spans from 23.03 million to 2.58 million years ago. It features 2 epochs: the Miocene, and the Pliocene.

The Miocene epoch spans from 23.03 to 5.333 million years ago and is a period in which grass spread further, dominating a large portion of the world, at the expense of forests. Kelp forests evolved, encouraging the evolution of new species, such as sea otters. During this time, perissodactyla thrived, and evolved into many different varieties. Apes evolved into 30 species. The Tethys Sea finally closed with the creation of the Arabian Peninsula, leaving only remnants as the Black, Red, Mediterranean and Caspian Seas. This increased aridity. Many new plants evolved: 95% of modern seed plants evolved in the mid-Miocene.

The Pliocene epoch lasted from 5.333 to 2.58 million years ago. The Pliocene featured dramatic climactic changes, which ultimately led to modern species and plants. The Mediterranean Sea dried up for several million years (because the ice ages reduced sea levels, disconnecting the Atlantic from the Mediterranean, and evaporation rates exceeded inflow from rivers). *Australopithecus* evolved in Africa, beginning the human branch. The isthmus of Panama formed, and animals migrated between North and South America, wreaking havoc on local ecologies. Climatic changes brought: savannas that are still continuing to spread across the world; Indian monsoons; deserts in central Asia;

Figure 171: *Animals of the Miocene (Chalicotherium, Hyenadon, Entelodont...). Mammals are the dominant terrestrial vertebrates of the Cenozoic.*

and the beginnings of the Sahara desert. The world map has not changed much since, save for changes brought about by the glaciations of the Quaternary, such as the Great Lakes, Hudson Bay, and the Baltic sea.

Quaternary

The Quaternary spans from 2.58 million years ago to present day, and is the shortest geological period in the Phanerozoic Eon. It features modern animals, and dramatic changes in the climate. It is divided into two epochs: the Pleistocene and the Holocene.

The Pleistocene lasted from 2.58 million to 11,700 years ago. This epoch was marked by ice ages as a result of the cooling trend that started in the Mid-Eocene. There were at least four separate glaciation periods marked by the advance of ice caps as far south as 40 degrees N latitude in mountainous areas. Meanwhile, Africa experienced a trend of desiccation which resulted in the creation of the Sahara, Namib, and Kalahari deserts. Many animals evolved including mammoths, giant ground sloths, dire wolves, saber-toothed cats, and most famously *Homo sapiens*. 100,000 years ago marked the end of one of the worst droughts in Africa, and led to the expansion of primitive humans. As the Pleistocene drew to a close, a major extinction wiped out much of the world's

Figure 172: *Megafauna of the Pleistocene (mammoths, cave lions, woolly rhino, Megaloceros, horses)*

megafauna, including some of the hominid species, such as Neanderthals. All the continents were affected, but Africa to a lesser extent. It still retains many large animals, such as hippos.

The Holocene began 11,700 years ago and lasts to the present day. All recorded history and "the history of the world" lies within the boundaries of the Holocene epoch. Human activity is blamed for a mass extinction that began roughly 10,000 years ago, though the species becoming extinct have only been recorded since the Industrial Revolution. This is sometimes referred to as the "Sixth Extinction". Over 322 species have become extinct due to human activity since the Industrial Revolution.

Animal life

Early in the Cenozoic, following the K-Pg event, the planet was dominated by relatively small fauna, including small mammals, birds, reptiles, and amphibians. From a geological perspective, it did not take long for mammals and birds to greatly diversify in the absence of the dinosaurs that had dominated during the Mesozoic. Some flightless birds grew larger than humans. These species are sometimes referred to as "terror birds," and were formidable predators. Mammals came to occupy almost every available niche (both marine and terrestrial), and some also grew very large, attaining sizes not seen in most of today's terrestrial mammals.

Early animals were the Entelodon (a so-called "hell pig"), Paraceratherium (a hornless rhinoceros relative) and Basilosaurus (an early whale). The extinction of many large diapsid groups, such as flightless dinosaurs, Plesiosauria and

Pterosauria allowed mammals and birds to greatly diversify and become the world's predominant fauna.

Tectonics

Geologically, the Cenozoic is the era when the continents moved into their current positions. Australia-New Guinea, having split from Pangea during the early Cretaceous, drifted north and, eventually, collided with South-east Asia; Antarctica moved into its current position over the South Pole; the Atlantic Ocean widened and, later in the era (2.8 million years ago), South America became attached to North America with the isthmus of Panama.

India collided with Asia 55 to 45[366] million years ago creating the Himalayas; Arabia collided with Eurasia, closing the Tethys Ocean and creating the Zagros Mountains, around 35[367] million years ago.

The break-up of Gondwana in Late Cretaceous and Cenozoic times led to a shift in the river courses of various large African rivers including the Congo, Niger, Nile, Orange, Limpopo and Zambezi.

Climate

The Paleocene–Eocene Thermal Maximum at about 55.5[368] million years ago was a significant global warming event; however, since the Azolla event of 49[369] million years ago, the Cenozoic Era has been a period of long-term cooling. After the tectonic creation of Drake Passage at 41[370] million years ago, when South America fully detached from Antarctica during the Oligocene, the climate cooled significantly due to the advent of the Antarctic Circumpolar Current which brought cool deep Antarctic water to the surface. The cooling trend continued in the Miocene, with relatively short warmer periods. When South America became attached to North America creating the Isthmus of Panama around 2.8[371] million years ago, the Arctic region cooled due to the strengthening of the Humboldt and Gulf Stream currents, eventually leading to the glaciations of the Quaternary ice age, the current interglacial of which is the Holocene Epoch. Recent analysis of the geomagnetic reversal frequency, oxygen isotope record, and tectonic plate subduction rate, which are indicators of the changes in the heat flux at the core mantle boundary, climate and plate tectonic activity, shows that all these changes indicate similar rhythms on million years' timescale in the Cenozoic Era occurring with the common fundamental periodicity of ~13 Myr during most of the time.

Life

During the Cenozoic, mammals proliferated from a few small, simple, generalized forms into a diverse collection of terrestrial, marine, and flying animals, giving this period its other name, the Age of Mammals, despite the fact that there are more than twice as many bird species as mammal species. The Cenozoic is just as much the age of savannas, the age of co-dependent flowering plants and insects, and the age of birds. Grass also played a very important role in this era, shaping the evolution of the birds and mammals that fed on it. One group that diversified significantly in the Cenozoic as well were the snakes. Evolving in the Cenozoic, the variety of snakes increased tremendously, resulting in many colubrids, following the evolution of their current primary prey source, the rodents.

In the earlier part of the Cenozoic, the world was dominated by the gastornithid birds, terrestrial crocodiles like *Pristichampsus*, and a handful of primitive large mammal groups like uintatheres, mesonychids, and pantodonts. But as the forests began to recede and the climate began to cool, other mammals took over.

The Cenozoic is full of mammals both strange and familiar, including chalicotheres, creodonts, whales, primates, entelodonts, saber-toothed cats, mastodons and mammoths, three-toed horses, giant rhinoceros like *Indricotherium*, the rhinoceros-like brontotheres, various bizarre groups of mammals from South America, such as the vaguely elephant-like pyrotheres and the dog-like marsupial relatives called borhyaenids and the monotremes and marsupials of Australia.

Bibliography

* *British Caenozoic Fossils*, 1975, The Natural History Museum, London.
* *Geologic Time*, by Henry Roberts.
* *After the Dinosaurs: The Age of Mammals*, by Donald R. Prothero, Bloomington, Indiana: Indiana University Press, 2006. ISBN 978-0-253-34733-6.

External links

> Wikisourcehas the text of the 1911 *Encyclopædia Britannica*article *Cainozoic*.

> Wikimedia Commons has media related to *Cenozoic*.

- Western Australian Museum – The Age of the Mammals[372]

Paleogene

Paleogene Period	
66–23.03 million years ago	
PreЄЄ OSD C P T J K PgN	
Mean atmospheric O₂ content over period duration	c. 26 vol %[373,374] (130 % of modern level
Mean atmospheric CO₂ content over period duration	c. 500 ppm[375] (2 times pre-industrial level)
Mean surface temperature over period duration	c. 18 °C[376] (4 °C above modern level)

Key events in the Paleogene

```
35
30
40
45
50
60
65
```

An approximate timescale of key Paleogene events.
Axis scale: millions of years ago.

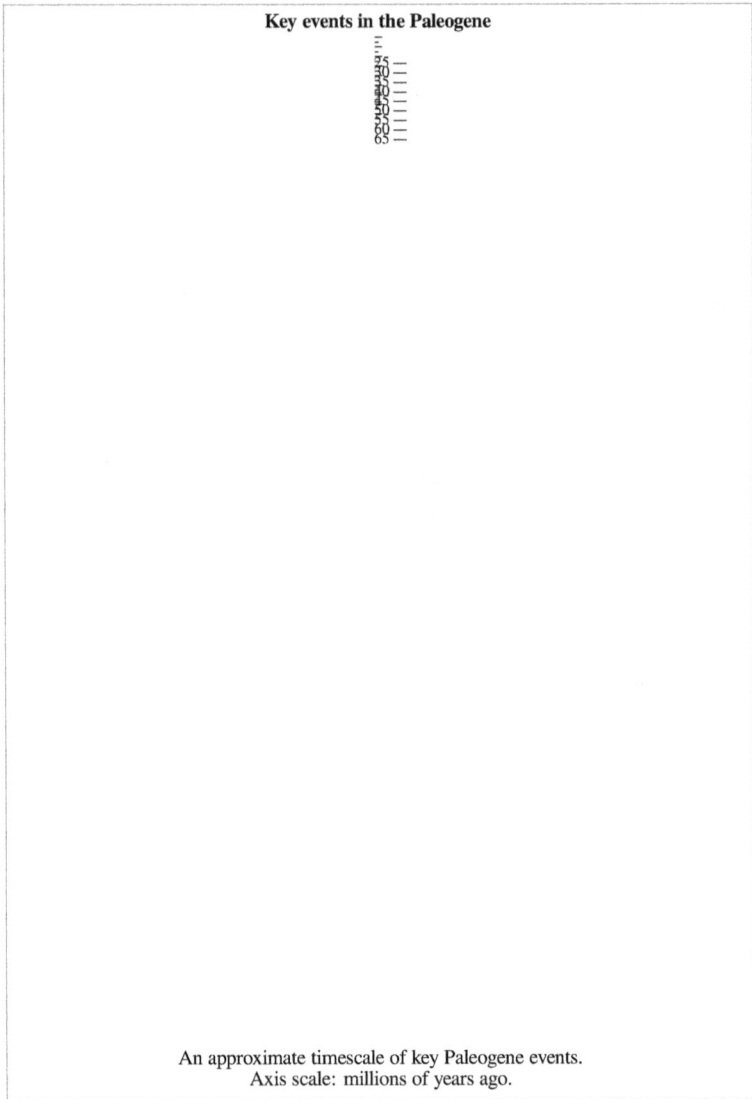

The **Paleogene** (/ˈpæliədʒiːn, <wbr />ˈpeɪliə-/; also spelled **Palaeogene** or
Palæogene; informally **Lower Tertiary** or **Early Tertiary**) is a geologic period and system that spans 43 million years from the end of the Cretaceous
Period 66 million years ago (Mya) to the beginning of the Neogene Period
23.03 Mya. It is the beginning of the Cenozoic Era of the present Phanerozoic Eon.[377] The Paleogene is most notable for being the time during which
mammals diversified from relatively small, simple forms into a large group of

diverse animals in the wake of the Cretaceous–Paleogene extinction event that ended the preceding Cretaceous Period.[378]

This period consists of the Paleocene, Eocene, and Oligocene epochs. The end of the Paleocene (55.5/54.8 Mya) was marked by the Paleocene–Eocene Thermal Maximum, one of the most significant periods of global change during the Cenozoic, which upset oceanic and atmospheric circulation and led to the extinction of numerous deep-sea benthic foraminifera and on land, a major turnover in mammals. The terms 'Paleogene System' (formal) and 'lower Tertiary System' (informal) are applied to the rocks deposited during the 'Paleogene Period'. The somewhat confusing terminology seems to be due to attempts to deal with the comparatively fine subdivisions of time possible in the relatively recent geologic past, for which more details are preserved. By dividing the Tertiary Period into two periods instead of directly into five epochs, the periods are more closely comparable to the duration of 'periods' of the preceding Mesozoic and Paleozoic Eras.

Climate and geography

The global climate during the Paleogene departed from the hot and humid conditions of the late Mesozoic era and began a cooling and drying trend which, despite having been periodically disrupted by warm periods such as the Paleocene–Eocene Thermal Maximum, persists today. The trend was partly caused by the formation of the Antarctic Circumpolar Current, which significantly lowered oceanic water temperatures.

During the Paleogene, the continents continued to drift closer to their current positions. India was in the process of colliding with Asia, subsequently forming the Himalayas. The Atlantic Ocean continued to widen by a few centimeters each year. Africa was moving north to meet with Europe and form the Mediterranean, while South America was moving closer to North America (they would later connect via the Isthmus of Panama). Inland seas retreated from North America early in the period. Australia had also separated from Antarctica and was drifting toward Southeast Asia.

Flora and fauna

Mammals began a rapid diversification during this period. After the Cretaceous–Paleogene extinction event, which saw the demise of the non-avian dinosaurs, mammals transformed from a few small and generalized forms that began to evolve into most of the modern varieties we see today. Some of these mammals would evolve into large forms that would dominate the land, while others would become capable of living in marine, specialized terrestrial, and airborne environments. Those that took to the oceans became modern cetaceans, while those that took to the trees became primates, the group to which humans belong. Birds, which were already well established by the end of the Cretaceous, also experienced an adaptive radiation as they took over the skies left empty by the now extinct Pterosaurs. In comparison to birds and mammals, most other branches of life remained relatively unchanged during this period.

As the Earth cooled, tropical plants became less numerous and were now restricted to equatorial regions. Deciduous plants, which could survive through the seasonal climates the world was now experiencing, became more common.

Geology

Oil industry relevance

The Paleogene is notable in the context of offshore oil drilling, and especially in Gulf of Mexico oil exploration, where it is commonly referred to as the "Lower Tertiary". These rock formations represent the current cutting edge of deep-water oil discovery.

Lower Tertiary rock formations encountered in the Gulf of Mexico oil industry usually tend to be comparatively high temperature and high pressure reservoirs, often with high sand content (70%+) or under very thick evaporite sediment layers.

Lower Tertiary explorations to date include (partial list):

- Kaskida Oil Field
- Tiber Oil Field
- Jack 2

External links

Wikisource has original works on the topic: *Cenozoic#Paleogene*

> Wikimedia Commons has media related to *Paleogene*.

- Paleogene Microfossils: 180+ images of Foraminifera[379]

Paleocene

System/ Period	Series/ Epoch	Stage/ Age	Age (Ma)	
Neogene	Miocene	Aquitanian	younger	
Paleogene	Oligocene	Chattian	23.03	27.82
		Rupelian	27.82	33.9
	Eocene	Priabonian	33.9	37.8
		Bartonian	37.8	41.2
		Lutetian	41.2	47.8
		Ypresian	47.8	56.0
	Paleocene	Thanetian	56.0	59.2
		Selandian	59.2	61.6
		Danian	61.6	66.0
Cretaceous	Upper/ Late	Maastrichtian	older	

Subdivision of the Paleogene Period according to the ICS, as of 2017.

The **Paleocene** (/'pæliəˌsiːn, <wbr />'pæ-, <wbr />-lioʊ-/) or **Palaeocene**, the "old recent", is a geological epoch that lasted from about 66 to 56[380] million years ago. It is the first epoch of the Paleogene Period in the modern Cenozoic Era. As with many geologic periods, the strata that define the epoch's beginning and end are well identified, but the exact ages remain uncertain.

The Paleocene Epoch is bracketed by two major events in Earth's history. It started with the mass extinction event at the end of the Cretaceous, known as the Cretaceous–Paleogene (K–Pg) boundary. This was a time marked by the demise of non-avian dinosaurs, giant marine reptiles and much other fauna and flora. The die-off of the dinosaurs left unfilled ecological niches worldwide. The Paleocene ended with the Paleocene–Eocene Thermal Maximum, a geologically brief (~0.2 million year) interval characterized by extreme changes in climate and carbon cycling.

The name "Paleocene" comes from Ancient Greek and refers to the "old(er)" (παλαιός, *palaios*) "new" (καινός, *kainos*) fauna that arose during the epoch.

Boundaries and subdivisions

The K–Pg boundary that marks the separation between Cretaceous and Paleocene is visible in the geological record of much of the Earth by a disconti-
nuity in the fossil fauna and high iridium levels. There is also fossil evidence
of abrupt changes in flora and fauna. There is some evidence that a substantial but very short-lived climatic change may have happened in the very early
decades of the Paleocene. There are several theories about the cause of the
K–Pg extinction event, with most evidence supporting the impact of a 10 km
diameter asteroid forming the buried Chicxulub crater on the coast of Yucatan,
Mexico.

The end of the Paleocene ($\approx$55.8 Ma) was also marked by a time of major change, one of the most significant periods of global change during the
Cenozoic. The Paleocene–Eocene Thermal Maximum upset oceanic and atmospheric circulation and led to the extinction of numerous deep-sea benthic
foraminifera and a major turnover in mammals on land.

The Paleocene is divided into three stages, the Danian, the Selandian and the
Thanetian, as shown in the table above. Additionally, the Paleocene is divided
into six Mammal Paleogene zones.

Climate

The early Paleocene was cooler and drier than the preceding Cretaceous,
though temperatures rose sharply during the Paleocene–Eocene Thermal Maximum. The climate became warm and humid worldwide towards the Eocene
boundary, with subtropical vegetation growing in Greenland and Patagonia,
crocodilians swimming off the coast of Greenland, and early primates evolving
in the tropical palm forests of northern Wyoming. The Earth's poles were cool
and temperate; North America, Europe, Australia and southern South America were warm and temperate; equatorial areas had tropical climates; and north
and south of the equatorial areas, climates were hot and arid, not dissimilar to
today's global desert belts around 30 degrees northern and southern latitude.

Paleogeography

In many ways, the Paleocene continued processes that had begun during the
late Cretaceous Period. During the Paleocene, the continents continued to
drift toward their present positions. Supercontinent Laurasia had not yet separated into three continents - Europe and Greenland were still connected,
North America and Asia were still intermittently joined by a land bridge, while
Greenland and North America were beginning to separate.[381] The Laramide

orogeny of the late Cretaceous continued to uplift the Rocky Mountains in the American west, which ended in the succeeding epoch.

South and North America remained separated by equatorial seas (they joined during the Neogene); the components of the former southern supercontinent Gondwanaland continued to split apart, with Africa, South America, Antarctica and Australia pulling away from each other. Africa was heading north towards Europe, slowly closing the Tethys Ocean, and India began its migration to Asia that would lead to a tectonic collision and the formation of the Himalayas.

The inland seas in North America (Western Interior Seaway) and Europe had receded by the beginning of the Paleocene, making way for new land-based flora and fauna.

Oceans

Warm seas circulated throughout the world, including the poles. The earliest Paleocene featured a low diversity and abundance of marine life, but this trend reversed later in the epoch. Tropical conditions gave rise to abundant marine life, including coral reefs. With the demise of marine reptiles at the end of the Cretaceous, sharks became the top predators. At the end of the Cretaceous, the ammonites and many species of foraminifera became extinct.

Marine fauna also came to resemble modern fauna, with only the marine mammals and the Carcharhinid sharks missing.

Flora

Terrestrial Paleocene strata immediately overlying the K–Pg boundary is in places marked by a "fern spike": a bed especially rich in fern fossils. Ferns are often the first species to colonize areas damaged by forest fires; thus the fern spike may indicate post-Chicxulub crater devastation.

In general, the Paleocene is marked by the development of modern plant species. Cacti and palm trees appeared. Paleocene and later plant fossils are generally attributed to modern genera or to closely related taxa.

The warm temperatures worldwide gave rise to thick tropical, sub-tropical and deciduous forest cover around the globe (the first recognizably modern rainforests) with ice-free polar regions covered with coniferous and deciduous trees. With no large browsing dinosaurs to thin them, Paleocene forests were probably denser than those of the Cretaceous.[382]

Flowering plants (angiosperms), first seen in the Cretaceous, continued to develop and proliferate, and along with them coevolved the insects that fed on these plants and pollinated them.

Figure 173: *Life restoration of Titanoides*

Fauna

Mammals

Mammals had first appeared in the Late Triassic, evolving from advanced cynodonts, and developed alongside the dinosaurs, exploiting ecological niches untouched by the larger and more famous Mesozoic animals: in the insect-rich forest underbrush and high up in the trees. These smaller mammals (as well as birds, reptiles, amphibians, and insects) survived the mass extinction at the end of the Cretaceous which wiped out the non-avian dinosaurs, and mammals diversified and spread throughout the world.

While early mammals were small nocturnal animals that mostly ate soft plant material and small animals such as insects, the demise of the non-avian dinosaurs and the beginning of the Paleocene saw mammals growing bigger and occupying a wider variety of ecological niches. Ten million years after the death of the non-avian dinosaurs, the world was filled with rodent-like mammals, medium-sized mammals scavenging in forests, and large herbivorous and carnivorous mammals hunting other mammals, birds, and reptiles.

Fossil evidence from the Paleocene is scarce, and there is relatively little known about mammals of the time. Because of their small size (constant until late in the epoch) early mammal bones are not well preserved in the fossil record, and most of what we know comes from fossil teeth (a much tougher substance), and only a few skeletons.

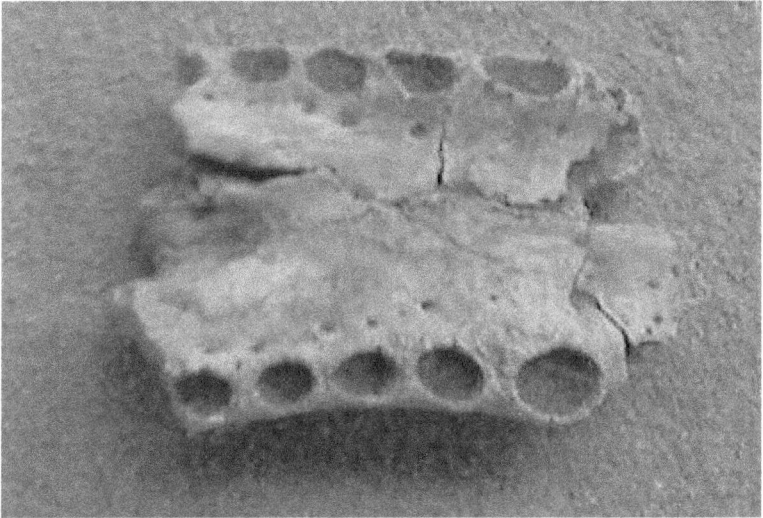

Figure 174: *Section of an Asiatosuchus jaw*

The brain to body mass ratios of these archaic mammals were quite low.[383]

Mammals of the Paleocene include:

- Monotremes: The ornithorhynchid *Obdurodon sudamericanum*, in the family that includes the platypus, is the only monotreme known from the Paleocene.
- Marsupials: modern kangaroos are marsupials, characterized by giving birth to embryonic young, who crawl into the mother's pouch and suckle until they are developed. The Bolivian *Pucadelphys andinus* and the North American *Peradectes* are two Paleocene examples.
- Multituberculates: the only major branch of mammals to become extinct since the K–Pg boundary, this rodent-like grouping includes the Paleocene *Ptilodus*.
- Placentals: this grouping of mammals became the most diverse and the most successful. Members include primates, plesiadapids, proboscideans, and hoofed ungulates, including the condylarths and the carnivorous mesonychids.

Reptiles

Because of the climatic conditions of the Paleocene, reptiles were more widely distributed over the globe than at present. Among the sub-tropical reptiles found in North America during this epoch are champsosaurs (fully aquatic

Figure 175: *Wannaganosuchus, a crocodilian from the Paleocene.*

reptiles), crocodilia, soft-shelled turtles, palaeophid snakes, varanid lizards, and *Protochelydra zangerli* (similar to modern snapping turtles).

Examples of champsosaurs of the Paleocene include *Champsosaurus gigas*, the largest champsosaur ever discovered. This creature was unusual among Paleocene non-squamate reptiles in that *C. gigas* became larger than its known Mesozoic ancestors: *C. gigas* is more than twice the length of the largest Cretaceous specimens (3 meters versus 1.5 meters). Another genus, *Simoedosaurus*, was similarly large; it appears rather suddenly in the fossil record, as its closest relatives occurred in the Early Cretaceous. Reptiles as a whole decreased in size after the K–Pg event. Champsosaurs declined towards the end of the Paleocene and became extinct during the Miocene.

Examples of Paleocene crocodylians are *Borealosuchus* (formerly *Leidyosuchus*) *formidabilis*, the apex predator and the largest animal of the Wannagan Creek fauna, and the alligatorid *Wannaganosuchus*.

Non-avian dinosaurs may have survived to some extent into the early Danian stage of the Paleocene Epoch circa 64.5 Mya. The controversial evidence for such is a hadrosaur leg bone found from Paleocene strata in New Mexico; but such stray late forms may be derived fossils.

Several species of snakes, such as *Titanoboa* and *Gigantophis*, grew to over 6 meters long.[384]

Birds

Birds began to re-diversify during the epoch, occupying new niches. Genetic studies suggest that nearly all modern bird clades can trace their origin to this epoch, with Neornithes having undergone an extremely fast, "star-like" radiation of species in the early Palaeocene in response to the vacancy of niches left by the KT event.[385]

Large flightless birds have been found in late Paleocene deposits, including the omnivorous *Gastornis* in Europe and carnivorous terror birds in South America, the latter of which survived until the Pleistocene.

Figure 176: *Gastornis*

In the late Paleocene, early owl types appeared, such as *Ogygoptynx* in the United States and *Berruornis* in France.

References

- Ogg, Jim (June 2004). "Overview of Global Boundary Stratotype Sections and Points (GSSP's)"[386]. Stratigraphy.org. Retrieved December 22, 2008.

External links

 Wikimedia Commons has media related to *Paleocene*.

> Wikisource has original works on the topic: *Cenozoic#Paleogene*

- Paleocene Mammals[387]
- BBC Changing Worlds: Paleocene[388]
- Maryland Paleocene Fossils[389]
- Paleos: Paleocene[390]
- Paleocene Evolutionary Radiation[391]
- PaleoMap Project[392]
- John Alroy, "Evidence of a Paleocene Evolutionary Radiation"[391]
- Paleocene Microfossils: 35+ images of Foraminifera[393]
- Petrified Wood Museum Palaeocene introduction[394]
- Smithsonian Paleocene Introduction[395]

Eocene

System/ Period	Series/ Epoch	Stage/ Age	Age (Ma)	
Neogene	Miocene	Aquitanian	younger	
Paleogene	Oligocene	Chattian	23.03	27.82
		Rupelian	27.82	33.9
	Eocene	Priabonian	33.9	37.8
		Bartonian	37.8	41.2
		Lutetian	41.2	47.8
		Ypresian	47.8	56.0
	Paleocene	Thanetian	56.0	59.2
		Selandian	59.2	61.6
		Danian	61.6	66.0
Cretaceous	Upper/ Late	Maastrichtian	older	

Subdivision of the Paleogene Period according to the ICS, as of 2017.

The **Eocene** (/ˈiːəˌsiːn, <wbr />ˈiːoʊ-/) Epoch, lasting from 56 to 33.9[396] million years ago, is a major division of the geologic timescale and the second epoch of the Paleogene Period in the Cenozoic Era. The Eocene spans the time from the end of the Paleocene Epoch to the beginning of the Oligocene Epoch. The start of the Eocene is marked by a brief period in which the concentration of the carbon isotope ^{13}C in the atmosphere was exceptionally low in comparison with the more common isotope ^{12}C. The end is set at a major extinction event called the *Grande Coupure* (the "Great Break" in continuity) or the Eocene–Oligocene extinction event, which may be related to the impact of one or more large bolides in Siberia and in what is now Chesapeake Bay. As with other geologic periods, the strata that define the start and end of the epoch are well identified,[397] though their exact dates are slightly uncertain.

The name *Eocene* comes from the Ancient Greek ἠώς (*ēós*, "dawn") and καινός (*kainós*, "new") and refers to the "dawn" of modern ('new') fauna that appeared during the epoch.[398]

Subdivisions

The Eocene epoch is conventionally divided into early, middle, and late subdivisions. The corresponding rocks are referred to as lower, middle, and upper Eocene. The Ypresian stage constitutes the lower, the Priabonian stage the upper; and the Lutetian and Bartonian stages are united as the middle Eocene.

Climate

The Eocene Epoch contained a wide variety of different climate conditions that includes the warmest climate in the Cenozoic Era and ends in an icehouse climate. The evolution of the Eocene climate began with warming after the end of the Palaeocene–Eocene Thermal Maximum (PETM) at 56 million years ago to a maximum during the Eocene Optimum at around 49 million years ago. During this period of time, little to no ice was present on Earth with a smaller difference in temperature from the equator to the poles. Following the maximum was a descent into an icehouse climate from the Eocene Optimum to the Eocene-Oligocene transition at 34 million years ago. During this decrease ice began to reappear at the poles, and the Eocene-Oligocene transition is the period of time where the Antarctic ice sheet began to rapidly expand.

Atmospheric greenhouse gas evolution

Greenhouse gases, in particular carbon dioxide and methane, played a significant role during the Eocene in controlling the surface temperature. The end of the PETM was met with a very large sequestration of carbon dioxide in the form of methane clathrate, coal, and crude oil at the bottom of the Arctic Ocean, that reduced the atmospheric carbon dioxide. This event was similar in magnitude to the massive release of greenhouse gasses at the beginning of the PETM, and it is hypothesized that the sequestration was mainly due to organic carbon burial and weathering of silicates. For the early Eocene there is much discussion on how much carbon dioxide was in the atmosphere. This is due to numerous proxies representing different atmospheric carbon dioxide content. For example, diverse geochemical and paleontological proxies indicate that at the maximum of global warmth the atmospheric carbon dioxide values were at 700–900 ppm while other proxies such as pedogenic (soil building) carbonate and marine boron isotopes indicate large changes of carbon dioxide of over 2,000 ppm over periods of time of less than 1 million years. Sources for this large influx of carbon dioxide could be attributed to volcanic out-gassing due to North Atlantic rifting or oxidation of methane stored in large reservoirs deposited from the PETM event in the sea floor or wetland environments. For contrast, today the carbon dioxide levels are at 400 ppm or 0.04%.

At about the beginning of the Eocene Epoch (55.8–33.9 million years ago) the amount of oxygen in the earth's atmosphere more or less doubled.[399]

During the early Eocene, methane was another greenhouse gas that had a drastic effect on the climate. In comparison to carbon dioxide, methane has much greater effect on temperature as methane is around 34 times more effective per molecule than carbon dioxide on a 100-year scale (it has a higher global warming potential).[400] Most of the methane released to the atmosphere during this period of time would have been from wetlands, swamps, and forests. The atmospheric methane concentration today is 0.000179% or 1.79 ppmv. Due to the warmer climate and sea level rise associated with the early Eocene, more wetlands, more forests, and more coal deposits would be available for methane release. Comparing the early Eocene production of methane to current levels of atmospheric methane, the early Eocene would be able to produce triple the amount of current methane production. The warm temperatures during the early Eocene could have increased methane production rates, and methane that is released into the atmosphere would in turn warm the troposphere, cool the stratosphere, and produce water vapor and carbon dioxide through oxidation. Biogenic production of methane produces carbon dioxide and water vapor along with the methane, as well as yielding infrared radiation. The breakdown of methane in an oxygen atmosphere produces carbon monoxide, water vapor and infrared radiation. The carbon monoxide is not stable so it eventually becomes carbon dioxide and in doing so releases yet more infrared radiation. Water vapor traps more infrared than does carbon dioxide.

The middle to late Eocene marks not only the switch from warming to cooling, but also the change in carbon dioxide from increasing to decreasing. At the end of the Eocene Optimum, carbon dioxide began decreasing due to increased siliceous plankton productivity and marine carbon burial. At the beginning of the middle Eocene an event that may have triggered or helped with the draw down of carbon dioxide was the Azolla event at around 49 million years ago. With the equable climate during the early Eocene, warm temperatures in the arctic allowed for the growth of azolla, which is a floating aquatic fern, on the Arctic Ocean. Compared to current carbon dioxide levels, these azolla grew rapidly in the enhanced carbon dioxide levels found in the early Eocene. As these azolla sank into the Arctic Ocean, they became buried and sequestered their carbon into the seabed. This event could have led to a draw down of atmospheric carbon dioxide of up to 470 ppm. Assuming the carbon dioxide concentrations were at 900 ppmv prior to the Azolla Event they would have dropped to 430 ppmv, or 30 ppmv more than they are today, after the Azolla Event. Another event during the middle Eocene that was a sudden and temporary reversal of the cooling conditions was the Middle Eocene Climatic Optimum. At around 41.5 million years ago, stable isotopic analysis of samples from Southern Ocean drilling sites indicated a warming event for 600 thousand years. A sharp increase in atmospheric carbon dioxide was observed with a maximum of 4000 ppm: the highest amount of atmospheric carbon dioxide detected during the Eocene. The main hypothesis for such a radical transition was due to the continental drift and collision of the India continent with the Asia continent and the resulting formation of the Himalayas. Another hypothesis involves extensive sea floor rifting and metamorphic decarbonation reactions releasing considerable amounts of carbon dioxide to the atmosphere.

At the end of the Middle Eocene Climatic Optimum, cooling and the carbon dioxide drawdown continued through the late Eocene and into the Eocene-Oligocene transition around 34 million years ago. Multiple proxies, such as oxygen isotopes and alkenones, indicate that at the Eocene-Oligocene transition, the atmospheric carbon dioxide concentration had decreased to around 750–800 ppm, approximately twice that of present levels.

Early Eocene and the equable climate problem

One of the unique features of the Eocene's climate as mentioned before was the equable and homogeneous climate that existed in the early parts of the Eocene. A multitude of proxies support the presence of a warmer equable climate being present during this period of time. A few of these proxies include the presence of fossils native to warm climates, such as crocodiles, located in the higher latitudes, the presence in the high latitudes of frost-intolerant flora such as palm trees which cannot survive during sustained freezes, and fossils of snakes found in the tropics that would require much higher average temperatures to sustain them. Using isotope proxies to determine ocean temperatures indicates sea surface temperatures in the tropics as high as 35 °C (95 °F) and, relative to present-day values, bottom water temperatures that are 10°C (18°F) higher. With these bottom water temperatures, temperatures in areas where deep-water forms near the poles are unable to be much cooler than the bottom water temperatures.

An issue arises, however, when trying to model the Eocene and reproduce the results that are found with the proxy data. Using all different ranges of greenhouse gasses that occurred during the early Eocene, models were unable to produce the warming that was found at the poles and the reduced seasonality that occurs with winters at the poles being substantially warmer. The models, while accurately predicting the tropics, tend to produce significantly cooler temperatures of up to 20°C (36°F) colder than the actual determined temperature at the poles. This error has been classified as the "equable climate problem". To solve this problem, the solution would involve finding a process to warm the poles without warming the tropics. Some hypotheses and tests which attempt to find the process are listed below.

Large lakes

Due to the nature of water as opposed to land, less temperature variability would be present if a large body of water is also present. In an attempt to try to mitigate the cooling polar temperatures, large lakes were proposed to mitigate seasonal climate changes. To replicate this case, a lake was inserted into North America and a climate model was run using varying carbon dioxide levels. The model runs concluded that while the lake did reduce the seasonality of the region greater than just an increase in carbon dioxide, the addition of a large lake was unable to reduce the seasonality to the levels shown by the floral and faunal data.

Ocean heat transport

The transport of heat from the tropics to the poles, much like how ocean heat transport functions in modern times, was considered a possibility for the increased temperature and reduced seasonality for the poles. With the increased sea surface temperatures and the increased temperature of the deep ocean water during the early Eocene, one common hypothesis was that due to these increases there would be a greater transport of heat from the tropics to the poles. Simulating these differences, the models produced lower heat transport due to the lower temperature gradients and were unsuccessful in producing an equable climate from only ocean heat transport.

Orbital parameters

While typically seen as a control on ice growth and seasonality, the orbital parameters were theorized as a possible control on continental temperatures and seasonality. Simulating the Eocene by using an ice free planet, eccentricity, obliquity, and precession were modified in different model runs to determine all the possible different scenarios that could occur and their effects on temperature. One particular case led to warmer winters and cooler summer by up to 30% in the North American continent, and it reduced the seasonal variation of temperature by up to 75%. While orbital parameters did not produce the warming at the poles, the parameters did show a great effect on seasonality and needed to be considered.

Polar stratospheric clouds

Another method considered for producing the warm polar temperatures were polar stratospheric clouds. Polar stratospheric clouds are clouds that occur in the lower stratosphere at very low temperatures. Polar stratospheric clouds have a great impact on radiative forcing. Due to their minimal albedo properties and their optical thickness, polar stratospheric clouds act similar to a greenhouse gas and traps outgoing longwave radiation. Different types of polar stratospheric clouds occur in the atmosphere: polar stratospheric clouds that are created due to interactions with nitric or sulfuric acid and water (Type I) or polar stratospheric clouds that are created with only water ice (Type II).

Methane is an important factor in the creation of the primary Type II polar stratospheric clouds that were created in the early Eocene. Since water vapor is the only supporting substance used in Type II polar stratospheric clouds, the presence of water vapor in the lower stratosphere is necessary where in most situations the presence of water vapor in the lower stratosphere is rare. When methane is oxidized, a significant amount of water vapor is released. Another requirement for polar stratospheric clouds is cold temperatures to ensure condensation and cloud production. Polar stratospheric cloud production, since it requires the cold temperatures, is usually limited to nighttime and winter conditions. With this combination of wetter and colder conditions in the lower stratosphere, polar stratospheric clouds could have formed over wide areas in Polar Regions.

To test the polar stratospheric clouds effects on the Eocene climate, models were run comparing the effects of polar stratospheric clouds at the poles to an increase in atmospheric carbon dioxide. The polar stratospheric clouds had a warming effect on the poles, increasing temperatures by up to 20 °C in the winter months. A multitude of feedbacks also occurred in the models due to the polar stratospheric clouds' presence. Any ice growth was slowed immensely and would lead to any present ice melting. Only the poles were affected with the change in temperature and the tropics were unaffected, which with an increase in atmospheric carbon dioxide would also cause the tropics to increase in temperature. Due to the warming of the troposphere from the increased greenhouse effect of the polar stratospheric clouds, the stratosphere would cool and would potentially increase the amount of polar stratospheric clouds.

While the polar stratospheric clouds could explain the reduction of the equator to pole temperature gradient and the increased temperatures at the poles during the early Eocene, there are a few drawbacks to maintaining polar stratospheric clouds for an extended period of time. Separate model runs were used to determine the sustainability of the polar stratospheric clouds. Methane would need to be continually released and sustained to maintain the lower stratospheric water vapor. Increasing amounts of ice and condensation nuclei would be need to be high for the polar stratospheric cloud to sustain itself and eventually expand.

Hyperthermals through the Early Eocene

During the warming in the Early Eocene between 52 and 55 million years ago, there were a series of short-term changes of carbon isotope composition in the ocean. These isotope changes occurred due to the release of carbon from the ocean into the atmosphere that led to a temperature increase of 4-8°C (7-14°F) at the surface of the ocean. These hyperthermals led to increased perturbations in planktonic and benthic foraminifera, with a higher rate of sedimentation as a consequence of the warmer temperatures. Recent analysis of and research into these hyperthermals in the early Eocene has led to hypotheses that the hyperthermals are based on orbital parameters, in particular eccentricity and obliquity. The hyperthermals in the early Eocene, notably the Palaeocene–Eocene Thermal Maximum (PETM), the Eocene Thermal Maximum 2 (ETM2), and the Eocene Thermal Maximum 3 (ETM3), were analyzed and found that orbital control may have had a role in triggering the ETM2 and ETM3.

Greenhouse to icehouse climate

The Eocene is not only known for containing the warmest period during the Cenozoic, but it also marked the decline into an icehouse climate and the rapid expansion of the Antarctic ice sheet. The transition from a warming climate into a cooling climate began at ~49 million years ago. Isotopes of carbon and oxygen indicate a shift to a global cooling climate. The cause of the cooling has been attributed to a significant decrease of >2000 ppm in atmospheric carbon dioxide concentrations. One proposed cause of the reduction in carbon dioxide during the warming to cooling transition was the azolla event. The increased warmth at the poles, the isolated Arctic basin during the early Eocene, and the significantly high amounts of carbon dioxide possibly led to azolla blooms across the Arctic Ocean. The isolation of the Arctic Ocean led to stagnant waters and as the azolla sank to the sea floor, they became part of the sediments and effectively sequestered the carbon. The ability for the azolla to sequester carbon is exceptional, and the enhanced burial of azolla could have had a significant effect on the world atmospheric carbon content and may have been the event to begin the transition into an ice house climate. Cooling after this event continued due to continual decrease in atmospheric carbon dioxide from organic productivity and weathering from mountain building.

Global cooling continued until there was a major reversal from cooling to warming indicated in the Southern Ocean at around 42–41 million years ago. Oxygen isotope analysis showed a large negative change in the proportion of heavier oxygen isotopes to lighter oxygen isotopes, which indicates an increase in global temperatures. This warming event is known as the Middle Eocene Climatic Optimum. The cause of the warming is considered to be primarily due to carbon dioxide increases, because carbon isotope signatures rule out major methane release during this short term warming. The increase in atmospheric carbon dioxide is considered to be due to increased seafloor spreading rates between Australia and Antarctica and increased amounts of volcanism in the region. Another possible atmospheric carbon dioxide increase could be during a sudden increase with metamorphic release during the Himalayan orogeny, however data on the exact timing of metamorphic release of atmospheric carbon dioxide is not well resolved in the data. Recent studies have mentioned, however, that the removal of the ocean between Asia and India could release significant amounts of carbon dioxide. This warming is short lived, as benthic oxygen isotope records indicate a return to cooling at ~40 million years ago.

Cooling continued throughout the rest of the late Eocene into the Eocene-Oligocene transition. During the cooling period, benthic oxygen isotopes show the possibility of ice creation and ice increase during this later cooling. The end of the Eocene and beginning of the Oligocene is marked with the massive expansion of area of the Antarctic ice sheet that was a major step into the icehouse climate. Along with the decrease of atmospheric carbon dioxide reducing the global temperature, orbital factors in ice creation can be seen with 100,000-year and 400,000-year fluctuations in benthic oxygen isotope records. Another major contribution to the expansion of the ice sheet was the creation of the Antarctic circumpolar current. The creation of the Antarctic circumpolar current would isolate the cold water around the Antarctic, which would reduce heat transport to the Antarctic along with creating ocean gyres that result in the upwelling of colder bottom waters. The issue with this hypothesis of the consideration of this being a factor for the Eocene-Oligocene transition is the timing of the creation of the circulation is uncertain. For Drake Passage, sediments indicate the opening occurred ~41 million years ago while tectonics indicate that this occurred ~32 million years ago.

Palaeogeography

During the Eocene, the continents continued to drift toward their present positions.

At the beginning of the period, Australia and Antarctica remained connected, and warm equatorial currents mixed with colder Antarctic waters, distributing the heat around the planet and keeping global temperatures high, but when Australia split from the southern continent around 45 Ma, the warm equatorial currents were routed away from Antarctica. An isolated cold water channel developed between the two continents. The Antarctic region cooled down, and the ocean surrounding Antarctica began to freeze, sending cold water and icefloes north, reinforcing the cooling.

The northern supercontinent of Laurasia began to fragment, as Europe, Greenland and North America drifted apart.

In western North America, mountain building started in the Eocene, and huge lakes formed in the high flat basins among uplifts, resulting in the deposition of the Green River Formation lagerstätte.

At about 35 Ma, an asteroid impact on the eastern coast of North America formed the Chesapeake Bay impact crater.

In Europe, the Tethys Sea finally disappeared, while the uplift of the Alps isolated its final remnant, the Mediterranean, and created another shallow sea with island archipelagos to the north. Though the North Atlantic was opening, a land connection appears to have remained between North America and Europe since the faunas of the two regions are very similar.

India began its collision with Asia, folding to initiate formation of the Himalayas.

It is hypothesized that the Eocene hothouse world was caused by runaway global warming from released methane clathrates deep in the oceans. The clathrates were buried beneath mud that was disturbed as the oceans warmed. Methane (CH_4) has ten to twenty times the greenhouse gas effect of carbon dioxide (CO_2).

Flora

At the beginning of the Eocene, the high temperatures and warm oceans created a moist, balmy environment, with forests spreading throughout the Earth from pole to pole. Apart from the driest deserts, Earth must have been entirely covered in forests.

Polar forests were quite extensive. Fossils and even preserved remains of trees such as swamp cypress and dawn redwood from the Eocene have been found on Ellesmere Island in the Arctic. Even at that time, Ellesmere Island was only a few degrees in latitude further south than it is today. Fossils of subtropical and even tropical trees and plants from the Eocene also have been found in Greenland and Alaska. Tropical rainforests grew as far north as northern North America and Europe.

Palm trees were growing as far north as Alaska and northern Europe during the early Eocene, although they became less abundant as the climate cooled. Dawn redwoods were far more extensive as well.

The earliest definitive Eucalyptus fossils were dated from 51.9 Mya, and were found in the Laguna del Hunco deposit in Chubut province in Argentina.

Cooling began mid-period, and by the end of the Eocene continental interiors had begun to dry out, with forests thinning out considerably in some areas. The newly evolved grasses were still confined to river banks and lake shores, and had not yet expanded into plains and savannas.

The cooling also brought seasonal changes. Deciduous trees, better able to cope with large temperature changes, began to overtake evergreen tropical species. By the end of the period, deciduous forests covered large parts of the northern continents, including North America, Eurasia and the Arctic, and rainforests held on only in equatorial South America, Africa, India and Australia.

Antarctica, which began the Eocene fringed with a warm temperate to subtropical rainforest, became much colder as the period progressed; the heat-loving tropical flora was wiped out, and by the beginning of the Oligocene, the continent hosted deciduous forests and vast stretches of tundra.

Fauna

The oldest known fossils of most of the modern mammal orders appear within a brief period during the early Eocene. At the beginning of the Eocene, several new mammal groups arrived in North America. These modern mammals, like artiodactyls, perissodactyls and primates, had features like long, thin legs, feet and hands capable of grasping, as well as differentiated teeth adapted for chewing. Dwarf forms reigned. All the members of the new mammal orders were small, under 10 kg; based on comparisons of tooth size, Eocene mammals were only 60% of the size of the primitive Palaeocene mammals that preceded them. They were also smaller than the mammals that followed them. It is assumed that the hot Eocene temperatures favored smaller animals that were better able to manage the heat.

Both groups of modern ungulates (hoofed animals) became prevalent because of a major radiation between Europe and North America, along with carnivorous ungulates like *Mesonyx*. Early forms of many other modern mammalian orders appeared, including bats, proboscidians (elephants), primates, rodents and marsupials. Older primitive forms of mammals declined in variety and importance. Important Eocene land fauna fossil remains have been found in

Figure 177: *Crassostrea gigantissima (Finch, 1824), a giant oyster from the Eocene of Texas*

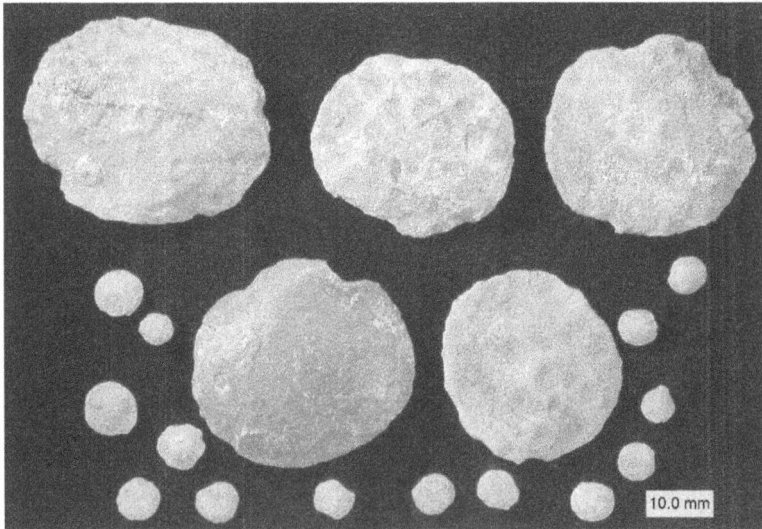

Figure 178: *Fossil nummulitid foraminiferans showing microspheric and megalospheric individuals; Eocene of the United Arab Emirates; scale in mm.*

Figure 179: *Basilosaurus*

western North America, Europe, Patagonia, Egypt and southeast Asia. Marine fauna are best known from South Asia and the southeast United States.

Reptile fossils from this time, such as fossils of pythons and turtles, are abundant. The remains of *Titanoboa*, a snake the length of a school bus, was discovered in South America along with other large reptilian megafauna. During the Eocene, plants and marine faunas became quite modern. Many modern bird orders first appeared in the Eocene.

Several rich fossil insect faunas are known from the Eocene, notably the Baltic amber found mainly along the south coast of the Baltic Sea, amber from the Paris Basin, France, the Fur Formation, Denmark and the Bembridge Marls from the Isle of Wight, England. Insects found in Eocene deposits are mostly assignable to modern genera, though frequently these genera do not occur in the area at present. For instance the bibionid genus *Plecia* is common in fossil faunas from presently temperate areas, but only lives in the tropics and subtropics today.

Oceans

The Eocene oceans were warm and teeming with fish and other sea life. The first carcharinid sharks evolved, as did early marine mammals, including *Basilosaurus*, an early species of whale that is thought to be descended from land animals that existed earlier in the Eocene. The first sirenians, relatives of the elephants, also evolved at this time.

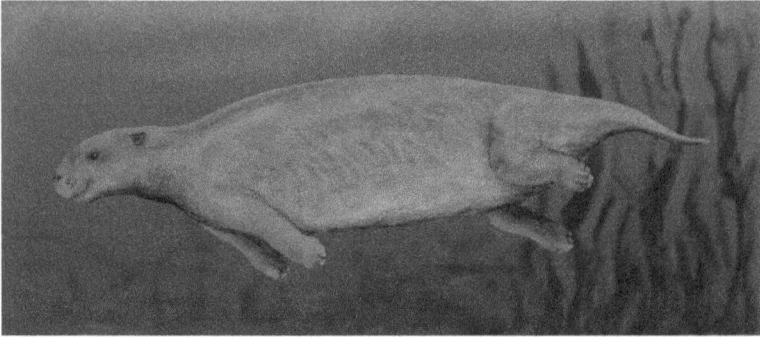

Figure 180: *Prorastomus, an early sirenian*

Eocene–Oligocene extinction

The end of the Eocene was marked by the Eocene–Oligocene extinction event, also known as the *Grande Coupure*.

Further reading

- Ogg, Jim; June, 2004, *Overview of Global Boundary Stratotype Sections and Points (GSSP's)* https://web.archive.org/web/20060716071827/http://www.stratigraphy.org/gssp.htm Accessed April 30, 2006.
- Stanley, Steven M. *Earth System History.* New York: W.H. Freeman and Company, 1999. ISBN 0-7167-2882-6

External links

Wikimedia Commons has media related to *Eocene*.

Wikisource has original works on the topic: *Cenozoic#Paleogene*

- PaleoMap Project[401]
- Paleos Eocene page[402]
- PBS Deep Time: Eocene[403]
- Eocene and Oligocene Fossils[404]
- The UPenn Fossil Forest Project, focusing on the Eocene polar forests in Ellesmere Island, Canada[405]
- Basilosaurus Primitive Eocene Whales[406]
- Basilosaurus - The plesiosaur that wasn't....[407]
- Detailed maps of Tertiary Western North America[408]
- Map of Eocene Earth[409]
- Eocene Microfossils: 60+ images of Foraminifera[410]
- Eocene Epoch. (2011). In Encyclopædia Britannica. Retrieved from http://www.britannica.com/EBchecked/topic/189322/Eocene-Epoch

Oligocene

System/ Period	Series/ Epoch	Stage/ Age	Age (Ma)	
Neogene	Miocene	Aquitanian	younger	
Paleogene	Oligocene	Chattian	23.03	27.82
		Rupelian	27.82	33.9
	Eocene	Priabonian	33.9	37.8
		Bartonian	37.8	41.2
		Lutetian	41.2	47.8
		Ypresian	47.8	56.0
	Paleocene	Thanetian	56.0	59.2
		Selandian	59.2	61.6
		Danian	61.6	66.0
Cretaceous	Upper/ Late	Maastrichtian	older	

Subdivision of the Paleogene Period according to the ICS, as of 2017.

The **Oligocene** (/'ɒlɪɡoʊsiːn/) is a geologic epoch of the Paleogene Period and extends from about 33.9 million to 23 million years before the present (33.9±0.1 to 23.03±0.05 Ma). As with other older geologic periods, the rock beds that define the epoch are well identified but the exact dates of the start and end of the epoch are slightly uncertain. The name Oligocene comes from the Ancient Greek ὀλίγος (*olígos*, "few") and καινός (*kainós*, "new"), and refers to the sparsity of extant forms of molluscs. The Oligocene is preceded by the Eocene Epoch and is followed by the Miocene Epoch. The Oligocene is the third and final epoch of the Paleogene Period.

The Oligocene is often considered an important time of transition, a link between the archaic world of the tropical Eocene and the more modern ecosystems of the Miocene.[411] Major changes during the Oligocene included a global expansion of grasslands, and a regression of tropical broad leaf forests to the equatorial belt.

The start of the Oligocene is marked by a notable extinction event called the Grande Coupure; it featured the replacement of European fauna with Asian fauna, except for the endemic rodent and marsupial families. By contrast, the Oligocene–Miocene boundary is not set at an easily identified worldwide event but rather at regional boundaries between the warmer late Oligocene and the relatively cooler Miocene.

Subdivisions

Oligocene faunal stages from youngest to oldest are:

Chattian or late Oligocene	(28.1[412] – 23.03[413] mya)
Rupelian or early Oligocene	(33.9[414] – 28.1[412] mya)

Climate

The Paleogene Period general temperature decline is interrupted by an Oligocene 7-million-year stepwise climate change. A deeper 8.2 °C, 400,000-year temperature depression leads the 2 °C, seven-million-year stepwise climate change 33.5 Ma (million years ago).[415,416] The stepwise climate change began 32.5 Ma and lasted through to 25.5 Ma, as depicted in the PaleoTemps chart. The Oligocene climate change was a global[417] increase in ice volume and a 55 m (181 feet) decrease in sea level (35.7–33.5 Ma) with a closely related (25.5–32.5 Ma) temperature depression.[418] The 7-million-year depression abruptly terminated within 1–2 million years of the La Garita Caldera eruption at 28–26 Ma. A deep 400,000-year glaciated Oligocene Miocene boundary event is recorded at McMurdo Sound and King George Island.

Paleogeography

During this epoch, the continents continued to drift toward their present positions. Antarctica became more isolated and finally developed an ice cap.

Mountain building in western North America continued, and the Alps started to rise in Europe as the African plate continued to push north into the Eurasian plate, isolating the remnants of the Tethys Sea. A brief marine incursion marks the early Oligocene in Europe. Marine fossils from the Oligocene are rare in North America. There appears to have been a land bridge in the early Oligocene between North America and Europe, since the faunas of the two regions are very similar. Sometime during the Oligocene, South America was finally detached from Antarctica and drifted north towards North America. It also allowed the Antarctic Circumpolar Current to flow, rapidly cooling the Antarctic continent.

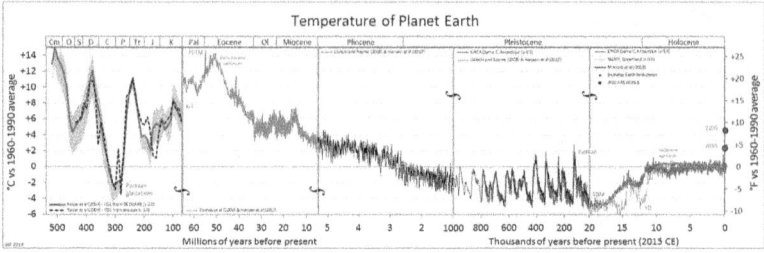

Figure 181: *Neotethys in oligocene (Rupelian, 33,9 — 28,4 mya)*

Flora

Angiosperms continued their expansion throughout the world as tropical and sub-tropical forests were replaced by temperate deciduous forests. Open plains and deserts became more common and grasses expanded from their water-bank habitat in the Eocene moving out into open tracts. However, even at the end of the period, grass was not quite common enough for modern savannas.

In North America, subtropical species dominated with cashews and lychee trees present, and temperate trees such as roses,Wikipedia:Please clarify beeches, and pines were common. The legumes spread, while sedges, bul-rushes, and ferns continued their ascent.

Figure 182: *Restoration of Oligocene fauna of North America*

Fauna

Even more open landscapes allowed animals to grow to larger sizes than they had earlier in the Paleocene epoch 30 million years earlier. Marine faunas became fairly modern, as did terrestrial vertebrate fauna on the northern continents. This was probably more as a result of older forms dying out than as a result of more modern forms evolving. Many groups, such as equids, entelodonts, rhinos, merycoidodonts, and camelids, became more able to run during this time, adapting to the plains that were spreading as the Eocene rainforests receded. The first felid, *Proailurus*, originated in Asia during the late Oligocene and spread to Europe.

South America was isolated from the other continents and evolved a quite distinct fauna during the Oligocene. The South American continent became home to strange animals such as pyrotheres and astrapotheres, as well as litopterns and notoungulates. Sebecosuchians, terror birds, and carnivorous metatheres, like the borhyaenids remained the dominant predators.

Brontotheres died out in the Earliest Oligocene, and creodonts died out outside Africa and the Middle East at the end of the period. Multituberculates, an ancient lineage of primitive mammals that originated back in the Jurassic, also became extinct in the Oligocene, aside from the gondwanatheres. The

Figure 183: *Pyrotherium* romeroi with the notoungulate Rhynchippus equinus

Figure 184: *Reconstruction of Astrapotherium in natural habitat.*

Figure 185: *Macrauchenia: a Litoptern*

Oligocene was home to a wide variety of strange mammals. A good example of this would be the White River Fauna of central North America, which were formerly a semiarid prairie home to many different types of endemic mammals, including entelodonts like *Archaeotherium*, camelids (such as *Poebrotherium*), running rhinoceratoids, three-toed equids (such as *Mesohippus*), nimravids, protoceratids, and early canids like *Hesperocyon*. Merycoidodonts, an endemic American group, were very diverse during this time. In Asia during the Oligocene, a group of running rhinoceratoids gave rise to the indricotheres, like *Paraceratherium*, which were the largest land mammals ever to walk the Earth.

The marine animals of Oligocene oceans resembled today's fauna, such as the bivalves. Calcareous cirratulids appeared in the Oligocene. The fossil record of marine mammals is a little spotty during this time, and not as well known as the Eocene or Miocene, but some fossils have been found. The baleen whales and toothed whales had just appeared, and their ancestors, the archaeocete cetaceans began to decrease in diversity due to their lack of echolocation, which was very useful as the water became colder and cloudier. Other factors to their decline could include climate changes and competition with today's modern cetaceans and the carcharhinid sharks, which also appeared in this epoch. Early desmostylians, like *Behemotops*, are known from the Oligocene. Pinnipeds appeared near the end of the epoch from an otter-like ancestor.

Oceans

The Oligocene sees the beginnings of modern ocean circulation, with tectonic shifts causing the opening and closing of ocean gateways. Cooling of the oceans had already commenced by the Eocene/Oligocene boundary, and they continued to cool as the Oligocene progressed. The formation of permanent Antarctic ice sheets during the early Oligocene and possible glacial activity in the Arctic may have influenced this oceanic cooling, though the extent of this influence is still a matter of some significant dispute.

The effects of oceanic gateways on circulation

The opening and closing of ocean gateways: the opening of the Drake Passage; the opening of the Tasmanian Gateway and the closing of the Tethys seaway; along with the final formation of the Greenland–Iceland–Faroes sill; played vital parts in reshaping oceanic currents during the Oligocene. As the continents shifted to a more modern configuration, so too did ocean circulation.

The Drake Passage

The Drake Passage is located between South America and Antarctica. Once the Tasmanian Gateway between Australia and Antarctica opened, all that kept Antarctica from being completely isolated by the Southern Ocean was its connection to South America. As the South American continent moved north, the Drake Passage opened and enabled the formation of the Antarctic Circumpolar Current (ACC), which would have kept the cold waters of Antarctica circulating around that continent and strengthened the formation of Antarctic Bottom Water (ABW). With the cold water concentrated around Antarctica, sea surface temperatures and, consequently, continental temperatures would have dropped. The onset of Antarctic glaciation occurred during the early Oligocene, and the effect of the Drake Passage opening on this glaciation has been the subject of much research. However, some controversy still exists as to the exact timing of the passage opening, whether it occurred at the start of the Oligocene or nearer the end. Even so, many theories agree that at the Eocene/Oligocene (E/O) boundary, a yet shallow flow existed between South America and Antarctica, permitting the start of an Antarctic Circumpolar Current.

Stemming from the issue of when the opening of the Drake Passage took place, is the dispute over how great of an influence the opening of the Drake Passage had on the global climate. While early researchers concluded that the advent of the ACC was highly important, perhaps even the trigger, for Antarctic glaciation and subsequent global cooling, other studies have suggested that the $\delta^{18}O$ signature is too strong for glaciation to be the main trigger for cooling.

Through study of Pacific Ocean sediments, other researchers have shown that the transition from warm Eocene ocean temperatures to cool Oligocene ocean temperatures took only 300,000 years, which strongly implies that feedbacks and factors other than the ACC were integral to the rapid cooling.

The late Oligocene opening of the Drake Passage

The latest hypothesized time for the opening of the Drake Passage is during the early Miocene. Despite the shallow flow between South America and Antarctica, there was not enough of a deep water opening to allow for significant flow to create a true Antarctic Circumpolar Current. If the opening occurred as late as hypothesized, then the Antarctic Circumpolar Current could not have had much of an effect on early Oligocene cooling, as it would not have existed.

The early Oligocene opening of the Drake Passage

The earliest hypothesized time for the opening of the Drake Passage is around 30 Ma. One of the possible issues with this timing was the continental debris cluttering up the seaway between the two plates in question. This debris, along with what is known as the Shackleton Fracture Zone, has been shown in a recent study to be fairly young, only about 8 million years old. The study concludes that the Drake Passage would be free to allow significant deep water flow by around 31 Ma. This would have facilitated an earlier onset of the Antarctic Circumpolar Current.

Currently, an opening of the Drake Passage during the early Oligocene is favored.

The opening of the Tasman Gateway

The other major oceanic gateway opening during this time was the Tasman, or Tasmanian, depending on the paper, gateway between Australia and Antarctica. The time frame for this opening is less disputed than the Drake Passage and is largely considered to have occurred around 34 Ma. As the gateway widened, the Antarctic Circumpolar Current strengthened.

The Tethys Seaway closing

The Tethys Seaway was not a gateway, but rather a sea in its own right. Its closing during the Oligocene had significant impact on both ocean circulation and climate. The collisions of the African plate with the European plate and of the Indian subcontinent with the Asian plate, cut off the Tethys Seaway that had provided a low-latitude ocean circulation. The closure of Tethys built some new mountains (the Zagros range) and drew down more carbon dioxide from the atmosphere, contributing to global cooling.

Greenland–Iceland–Faroes

The gradual separation of the clump of continental crust and the deepening of the tectonic sill in the North Atlantic that would become Greenland, Iceland, and the Faroe Islands helped to increase the deep water flow in that area. More information about the evolution of North Atlantic Deep Water will be given a few sections down.

Ocean cooling

Evidence for ocean-wide cooling during the Oligocene exists mostly in isotopic proxies. Patterns of extinction and patterns of species migration can also be studied to gain insight into ocean conditions. For a while, it was thought that the glaciation of Antarctica may have significantly contributed to the cooling of the ocean, however, recent evidence tends to deny this.

Deep water

Isotopic evidence suggests that during the early Oligocene, the main source of deep water was the North Pacific and the Southern Ocean. As the Greenland-Iceland-Faroe sill deepened and thereby connected the Norwegian–Greenland sea with the Atlantic Ocean, the deep water of the North Atlantic began to come into play as well. Computer models suggest that once this occurred, a more modern in appearance thermo-haline circulation started.

North Atlantic deep water

Evidence for the early Oligocene onset of chilled North Atlantic deep water lies in the beginnings of sediment drift deposition in the North Atlantic, such as the Feni and Southeast Faroe drifts.

South Ocean deep water

The chilling of the South Ocean deep water began in earnest once the Tasmanian Gateway and the Drake Passage opened fully. Regardless of the time at which the opening of the Drake Passage occurred, the effect on the cooling of the Southern Ocean would have been the same.

Impact events

Recorded extraterrestrial impacts:

- Haughton impact crater, Nunavut, Canada (23 Ma, crater 24 km (15 mi) diameter)

Supervolcanic explosions

La Garita Caldera (28 through 26 million years ago, VEI=9.2)

References

- Ogg, Jim; June, 2004, *Overview of Global Boundary Stratotype Sections and Points (GSSP's)* http://www.stratigraphy.org/gssp.htm Accessed April 30, 2006.

External links

Wikimedia Commons has media related to *Oligocene*.

Wikisource has original works on the topic: *Cenozoic#Paleogene*

- Palaeos: Oligocene[419]
- UCMP Berkeley Oligocene Page[420]
- Prehistoric Pictures, in the Public Domain[421]
- Oligocene Leaf Fossils[422]
- Olicgocene Fish Fossils[423]
- PaleoMap Project: Oligocene[424]
- Oligocene Microfossils: 300+ images of Foraminifera[425]

Neogene

Neogene Period	
23.03–2.58 million years ago	
PreЄЄ OSD C P T J K PgN	
Mean atmospheric O $_2$ content over period duration	c. 21.5 vol %[426,427] (108 % of modern level
Mean atmospheric CO $_2$ content over period duration	c. 280 ppm[428] (1 times pre-industrial level)
Mean surface tempera- ture over period duration	c. 14 °C[429] (0 °C above modern level)

Key events in the Neogene

0
2
3
5
6
8
10
12
14
16
18
20
24

An approximate timescale of key Neogene events.
Vertical axis: millions of years ago.

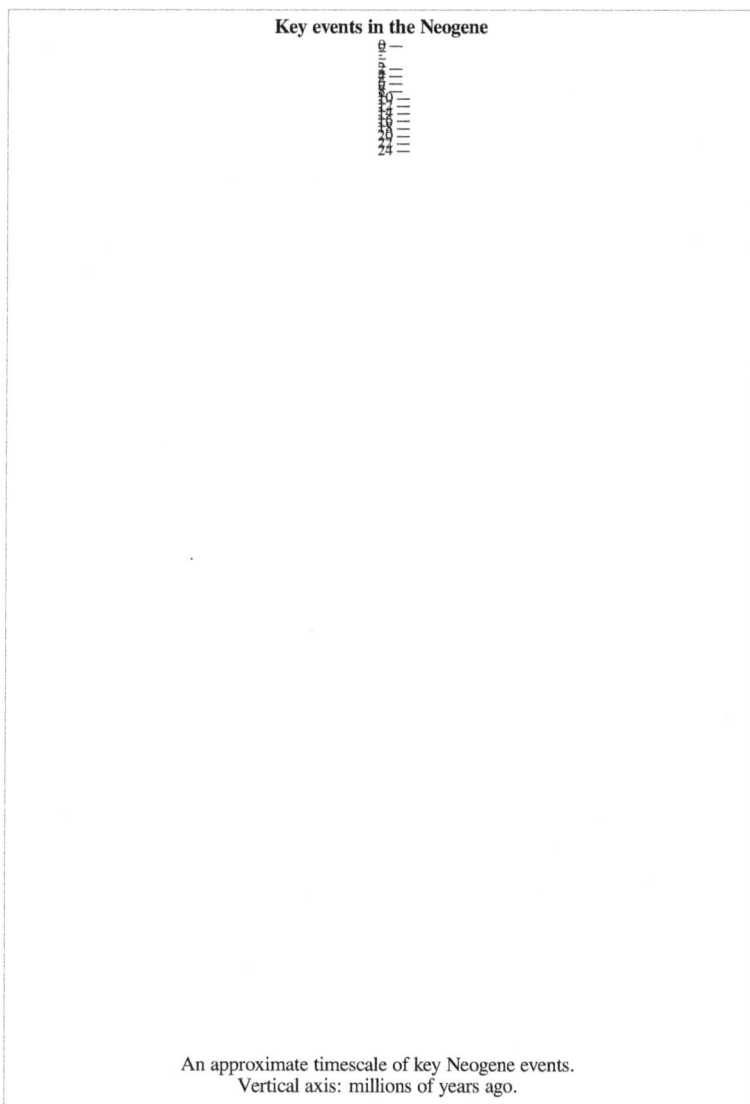

The **Neogene** (/ˈniːəˌdʒiːn/)[430] (informally **Upper Tertiary** or **Late Tertiary**) is a geologic period and system that spans 20.45 million years from the end of the Paleogene Period 23.03 million years ago (Mya) to the beginning of the present Quaternary Period 2.58 Mya. The Neogene is sub-divided into two epochs, the earlier Miocene and the later Pliocene. Some geologists assert that the Neogene cannot be clearly delineated from the modern geological period, the Quaternary.

During this period, mammals and birds continued to evolve into roughly modern forms, while other groups of life remained relatively unchanged. Early hominids, the ancestors of humans, appeared in Africa near the end of the period. Some continental movement took place, the most significant event being the connection of North and South America at the Isthmus of Panama, late in the Pliocene. This cut off the warm ocean currents from the Pacific to the Atlantic Ocean, leaving only the Gulf Stream to transfer heat to the Arctic Ocean. The global climate cooled considerably over the course of the Neogene, culminating in a series of continental glaciations in the Quaternary Period that follows.

Divisions

In ICS terminology, from upper (later, more recent) to lower (earlier):

The Pliocene Epoch is subdivided into 2 ages:

- Piacenzian Age, preceded by
- Zanclean Age

The Miocene Epoch is subdivided into 6 ages:

- Messinian Age, preceded by
- Tortonian Age
- Serravallian Age
- Langhian Age
- Burdigalian Age
- Aquitanian Age

In different geophysical regions of the world, other regional names are also used for the same or overlapping ages and other timeline subdivisions.

The terms *Neogene System* (formal) and *upper Tertiary System* (informal) describe the rocks deposited during the *Neogene Period*.

Geography

The continents in the Neogene were very close to their current positions. The Isthmus of Panama formed, connecting North and South America. The Indian subcontinent continued to collide with Asia, forming the Himalayas. Sea levels fell, creating land bridges between Africa and Eurasia and between Eurasia and North America.

Figure 186: *Scene featuring Miocene (Early Neogene) fauna*

Climate

The global climate became seasonal and continued an overall drying and cooling trend which began at the start of the Paleogene. The ice caps on both poles began to grow and thicken, and by the end of the period the first of a series of glaciations of the current Ice Age began.

Flora and fauna

Marine and continental flora and fauna have a modern appearance. The reptile group Choristodera became extinct in the early part of the period, while the amphibians known as Allocaudata disappeared at the end. Mammals and birds continued to be the dominant terrestrial vertebrates, and took many forms as they adapted to various habitats. The first hominins, the ancestors of humans, may have appeared in southern Europe and migrated into Africa.

In response to the cooler, seasonal climate, tropical plant species gave way to deciduous ones and grasslands replaced many forests. Grasses therefore greatly diversified, and herbivorous mammals evolved alongside it, creating the many grazing animals of today such as horses, antelope, and bison. Eucalyptus fossil leaves occur in the Miocene of New Zealand, where the genus is not native today, but have been introduced from Australia.

Disagreements

The Neogene traditionally ended at the end of the Pliocene Epoch, just before the older definition of the beginning of the Quaternary Period; many time scales show this division.

However, there was a movement amongst geologists (particularly marine geologists) to also include ongoing geological time (Quaternary) in the Neogene, while others (particularly terrestrial geologists) insist the Quaternary to be a separate period of distinctly different record. The somewhat confusing terminology and disagreement amongst geologists on where to draw what hierarchical boundaries is due to the comparatively fine divisibility of time units as time approaches the present, and due to geological preservation that causes the youngest sedimentary geological record to be preserved over a much larger area and to reflect many more environments than the older geological record.[431] By dividing the Cenozoic Era into three (arguably two) periods (Paleogene, Neogene, Quaternary) instead of seven epochs, the periods are more closely comparable to the duration of periods in the Mesozoic and Paleozoic eras.

The International Commission on Stratigraphy (ICS) once proposed that the Quaternary be considered a sub-era (sub-erathem) of the Neogene, with a beginning date of 2.58 Ma, namely the start of the Gelasian Stage. In the 2004 proposal of the ICS, the Neogene would have consisted of the Miocene and Pliocene epochs.[432] The International Union for Quaternary Research (INQUA) counterproposed that the Neogene and the Pliocene end at 2.58 Ma, that the Gelasian be transferred to the Pleistocene, and the Quaternary be recognized as the third period in the Cenozoic, citing key changes in Earth's climate, oceans, and biota that occurred 2.58 Ma and its correspondence to the Gauss-Matuyama magnetostratigraphic boundary.[433] In 2006 ICS and INQUA reached a compromise that made Quaternary a subera, subdividing Cenozoic into the old classical Tertiary and Quaternary, a compromise that was rejected by International Union of Geological Sciences because it split both Neogene and Pliocene in two.

Following formal discussions at the 2008 International Geological Congress in Oslo, Norway, the ICS decided in May 2009 to make the Quaternary the youngest period of the Cenozoic Era with its base at 2.58 Mya and including the Gelasian age, which was formerly considered part of the Neogene Period and Pliocene Epoch. Thus the Neogene Period ends bounding the succeeding Quaternary Period at 2.58 Mya.

External links

- Digital Atlas of Neogene Life for the Southeastern United States[434] — *by San Jose State University via Web Archive.*

Miocene

System/ Period	Series/ Epoch	Stage/ Age	Age (Ma)	
Quaternary	Pleistocene	Gelasian	younger	
Neogene	Pliocene	Piacenzian	2.58	3.600
		Zanclean	3.600	5.333
	Miocene	Messinian	5.333	7.246
		Tortonian	7.246	11.63
		Serravallian	11.63	13.82
		Langhian	13.82	15.97
		Burdigalian	15.97	20.44
		Aquitanian	20.44	23.03
Paleogene	Oligocene	Chattian	older	

Subdivision of the Neogene Period according to the ICS, as of 2017.

The **Miocene** (/ˈmaɪəˌsiːn/[435,436]) is the first geological epoch of the Neogene Period and extends from about 23.03 to 5.333[437] million years ago (Ma). The Miocene was named by Charles Lyell; its name comes from the Greek words μείων (*meiōn*, "less") and καινός (*kainos*, "new")[438] and means "less recent" because it has 18% fewer modern sea invertebrates than the Pliocene. The Miocene is preceded by the Oligocene and is followed by the Pliocene.

As the earth went from the Oligocene through the Miocene and into the Pliocene, the climate slowly cooled towards a series of ice ages. The Miocene boundaries are not marked by a single distinct global event but consist rather of regionally defined boundaries between the warmer Oligocene and the cooler Pliocene Epoch.

Apes arose and diversified during the Miocene, becoming widespread in the Old World. By the end of this epoch, the ancestors of humans had split away from the ancestors of the chimpanzees to follow their own evolutionary path (7.5 to 5.6[439] million years ago). As in the Oligocene before it, grasslands continued to expand and forests to dwindle in extent. In the seas of the Miocene, kelp forests made their first appearance and soon became one of Earth's most productive ecosystems.

The plants and animals of the Miocene were recognizably modern. Mammals and birds were well-established. Whales, pinnipeds, and kelp spread.

The Miocene is of particular interest to geologists and palaeoclimatologists as major phases of the geology of the Himalaya occurred during the Miocene, affecting monsoonal patterns in Asia, which were interlinked with glacial periods in the northern hemisphere.

Subdivisions

Hominin timeline

Axis scale: million years

🖐

Also see: *Life timeline* and *Nature timeline*

The Miocene faunal stages from youngest to oldest are typically named according to the International Commission on Stratigraphy:

Sub-epoch	Faunal stage	Time range
Late Miocene	Messinian	7.246–5.333 Ma
	Tortonian	11.608–7.246 Ma
Middle Miocene	Serravallian	13.65–11.608 Ma
	Langhian	15.97–13.65 Ma
Early Miocene	Burdigalian	20.43–15.97 Ma
	Aquitanian	23.03–20.43 Ma

Regionally, other systems are used, based on characteristic land mammals; some of them overlap with the preceding Oligocene and following Pliocene epochs:

European Land Mammal Ages

- Turolian (9.0 to 5.3 Ma)
- Vallesian (11.6 to 9.0 Ma)
- Astaracian (16.0 to 11.6 Ma)
- Orleanian (20.0 to 16.0 Ma)
- Agenian (23.8 to 20.0 Ma)

North American Land Mammal Ages

- Hemphillian (10.3 to 4.9 Ma)
- Clarendonian (13.6 to 10.3 Ma)
- Barstovian (16.3 to 13.6 Ma)
- Hemingfordian (20.6 to 16.3 Ma)
- Arikareean (30.6 to 20.6 Ma)

South American Land Mammal Ages

- Montehermosan (6.8 to 4.0 Ma)
- Huayquerian (9.0 to 6.8 Ma)
- Mayoan (11.8 to 9.0 Ma)
- Laventan (13.8 to 11.8 Ma)
- Colloncuran (15.5 to 13.8 Ma)
- Friasian (16.3 to 15.5 Ma)

- Santacrucian (17.5 to 16.3 Ma)
- Colhuehuapian (21.0 to 17.5 Ma)

Paleogeography

Continents continued to drift toward their present positions. Of the modern geologic features, only the land bridge between South America and North America was absent, although South America was approaching the western subduction zone in the Pacific Ocean, causing both the rise of the Andes and a southward extension of the Meso-American peninsula.

Mountain building took place in western North America, Europe, and East Asia. Both continental and marine Miocene deposits are common worldwide with marine outcrops common near modern shorelines. Well studied continental exposures occur in the North American Great Plains and in Argentina.

India continued to collide with Asia, creating dramatic new mountain ranges. The Tethys Seaway continued to shrink and then disappeared as Africa collided with Eurasia in the Turkish–Arabian region between 19 and 12 Ma. The subsequent uplift of mountains in the western Mediterranean region and a global fall in sea levels combined to cause a temporary drying up of the Mediterranean Sea (known as the Messinian salinity crisis) near the end of the Miocene.

The global trend was towards increasing aridity caused primarily by global cooling reducing the ability of the atmosphere to absorb moisture. Uplift of East Africa in the late Miocene was partly responsible for the shrinking of tropical rain forests in that region, and Australia got drier as it entered a zone of low rainfall in the Late Miocene.

South America

During the Oligocene and Early Miocene large swathes of Patagonia were subject to a marine transgression. The transgression might have temporarily linked the Pacific and Atlantic Oceans, as inferred from the findings of marine invertebrate fossils of both Atlantic and Pacific affinity in La Cascada Formation. Connection would have occurred through narrow epicontinental seaways that formed channels in a dissected topography. The Antarctic Plate started to subduct beneath South America 14 million years ago in the Miocene, forming the Chile Triple Junction. At first the Antarctic Plate subducted only in the southernmost tip of Patagonia, meaning that the Chile Triple Junction lay near the Strait of Magellan. As the southern part of Nazca Plate and the Chile Rise became consumed by subduction the more northerly regions of the Antarctic Plate begun to subduct beneath Patagonia so that the Chile Triple Junction advanced to the north over time. The asthenospheric window associated to the triple junction disturbed previous patterns of mantle convection beneath Patagonia inducing an uplift of ca. 1 km that reversed the Oligocene–Miocene transgression.

Climate

Climates remained moderately warm, although the slow global cooling that eventually led to the Pleistocene glaciations continued.

Although a long-term cooling trend was well underway, there is evidence of a warm period during the Miocene when the global climate rivalled that of the Oligocene. The Miocene warming began 21 million years ago and continued until 14 million years ago, when global temperatures took a sharp drop—the Middle Miocene Climate Transition (MMCT). By 8 million years ago, temperatures dropped sharply once again, and the Antarctic ice sheet was already approaching its present-day size and thickness. Greenland may have begun to have large glaciers as early as 7 to 8 million years ago,Wikipedia:Citation needed although the climate for the most part remained warm enough to support forests there well into the Pliocene.

Life

Life during the Miocene Epoch was mostly supported by the two newly formed biomes, kelp forests and grasslands. Grasslands allow for more grazers, such as horses, rhinoceroses, and hippos. Ninety five percent of modern plants existed by the end of this epoch.

Flora

The coevolution of gritty, fibrous, fire-tolerant grasses and long-legged gregarious ungulates with high-crowned teeth, led to a major expansion of grass-grazer ecosystems, with roaming herds of large, swift grazers pursued by predators across broad sweeps of open grasslands, displacing desert, woodland, and browsers.

The higher organic content and water retention of the deeper and richer grassland soils, with long term burial of carbon in sediments, produced a carbon and water vapor sink. This, combined with higher surface albedo and lower evapotranspiration of grassland, contributed to a cooler, drier climate. C_4 grasses, which are able to assimilate carbon dioxide and water more efficiently than C_3 grasses, expanded to become ecologically significant near the end of the Miocene between 6 and 7 million years ago. The expansion of grasslands and radiations among terrestrial herbivores correlates to fluctuations in CO_2.

Figure 187: *The dragon blood tree is considered a remnant of the Mio-Pliocene Laurasian subtropical forests that are now almost extinct in North Africa.*

Cycads between 11.5 and 5 m.y.a. began to rediversify after previous declines in variety due to climatic changes, and thus modern cycads are not a good model for a "living fossil". Eucalyptus fossil leaves occur in the Miocene of New Zealand, where the genus is not native today, but have been introduced from Australia.

Fauna

Both marine and continental fauna were fairly modern, although marine mammals were less numerous. Only in isolated South America and Australia did widely divergent fauna exist.

In the Early Miocene, several Oligocene groups were still diverse, including nimravids, entelodonts, and three-toed equids. Like in the previous Oligocene epoch, oreodonts were still diverse, only to disappear in the earliest Pliocene. During the later Miocene mammals were more modern, with easily recognizable canids, bears, procyonids, equids, beavers, deer, camelids, and whales, along with now extinct groups like borophagine canids, certain gomphotheres, three-toed horses, and semiaquatic and hornless rhinos like *Teleoceras* and *Aphelops*. Islands began to form between South and North America in the Late Miocene, allowing ground sloths like *Thinobadistes* to island-hop to North America. The expansion of silica-rich C_4 grasses led to worldwide extinctions of herbivorous species without high-crowned teeth.

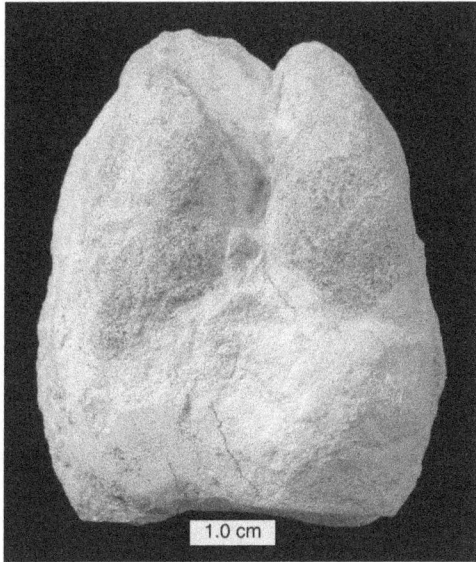

Figure 188: *Cameloid footprint (Lamaichnum alfi Sarjeant and Reynolds, 1999; convex hyporelief) from the Barstow Formation (Miocene) of Rainbow Basin, California.*

A few basal mammal groups endured into this epoch in southern landmasses, including the south american dryolestoid *Necrolestes* and gondwanathere *Patagonia* and New Zealand's Saint Bathans Mammal. Nonmarsupial metatherians were also still around, such as the American and Eurasian herpetotheriids and peradectids such as *Siamoperadectes*, and the South American sparassodonts.

Unequivocally recognizable dabbling ducks, plovers, typical owls, cockatoos and crows appear during the Miocene. By the epoch's end, all or almost all modern bird groups are believed to have been present; the few post-Miocene bird fossils which cannot be placed in the evolutionary tree with full confidence are simply too badly preserved, rather than too equivocal in character. Marine birds reached their highest diversity ever in the course of this epoch.

Approximately 100 species of apes lived during this time, ranging throughout Africa, Asia and Europe and varying widely in size, diet, and anatomy. Due to scanty fossil evidence it is unclear which ape or apes contributed to the modern hominid clade, but molecular evidence indicates this ape lived between 7 and 8 million years ago. The first hominins (bipedal apes of the human lineage) appeared in Africa at the very end of the Miocene, including

Figure 189: *Miocene fauna of North America, as restored by paleoartist Jay Matternes*

Sahelanthropus, Orrorin, and an early form of *Ardipithecus* (*A. kadabba*) The chimpanzee–human divergence is thought to have occurred at this time.

The expansion of grasslands in North America also led to an explosive radiation among snakes. Previously, snakes were a minor component of the North American fauna, but during the Miocene, the number of species and their prevalence increased dramatically with the first appearances of vipers and elapids in North America and the significant diversification of Colubridae (including the origin of many modern genera such as *Nerodia, Lampropeltis, Pituophis* and *Pantherophis*).

In the oceans, brown algae, called kelp, proliferated, supporting new species of sea life, including otters, fish and various invertebrates.

Cetaceans attained their greatest diversity during the Miocene, with over 20 recognized genera in comparison to only six living genera. This diversification correlates with emergence of gigantic macro-predators such as megatoothed sharks and raptorial sperm whales. Prominent examples are *C. megalodon* and *L. melvillei.* Other notable large sharks were *C. chubutensis, Isurus hastalis,* and *Hemipristis serra.*

Crocodilians also showed signs of diversification during Miocene. The largest form among them was a gigantic caiman *Purussaurus* which inhabited South

Figure 190: *Fossils from the Calvert Formation, Zone 10, Calvert Co., MD (Miocene).*

America. Another gigantic form was a false gharial *Rhamphosuchus*, which inhabited modern age India. A strange form, *Mourasuchus* also thrived alongside *Purussaurus*. This species developed a specialized filter-feeding mechanism, and it likely preyed upon small fauna despite its gigantic size.

The pinnipeds, which appeared near the end of the Oligocene, became more aquatic. Prominent genus was *Allodesmus*. A ferocious walrus, *Pelagiarctos* may have preyed upon other species of pinnipeds including *Allodesmus*.

Furthermore, South American waters witnessed the arrival of *Megapiranha paranensis*, which were considerably larger than modern age piranhas.

New Zealand's Miocene fossil record is particularly rich. Marine deposits showcase a variety of cetaceans and penguins, illustrating the evolution of both groups into modern representatives. The early Miocene Saint Bathans Fauna is the only Cenozoic terrestrial fossil record of the landmass, showcasing a wide variety of not only bird species, including early representatives of clades such as moas, kiwis and adzebills, but also a diverse herpetofauna of sphenodontians, crocodiles and turtle as well as a rich terrestrial mammal fauna composed of various species of bats and the enigmatic Saint Bathans Mammal.

Oceans

There is evidence from oxygen isotopes at Deep Sea Drilling Program sites that ice began to build up in Antarctica about 36 Ma during the Eocene. Further

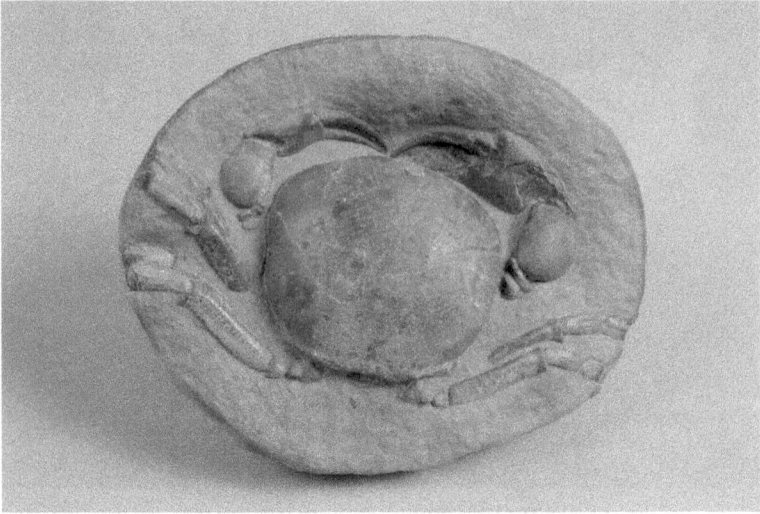

Figure 191: *A Miocene crab (Tumidocarcinus giganteus)*
from the collection of the Children's Museum of Indianapolis

marked decreases in temperature during the Middle Miocene at 15 Ma probably reflect increased ice growth in Antarctica. It can therefore be assumed that
East Antarctica had some glaciers during the early to mid Miocene (23–15
Ma). Oceans cooled partly due to the formation of the Antarctic Circumpolar
Current, and about 15 million years ago the ice cap in the southern hemisphere
started to grow to its present form. The Greenland ice cap developed later, in
the Middle Pliocene time, about 3 million years ago.

Middle Miocene disruption

The "Middle Miocene disruption" refers to a wave of extinctions of terrestrial
and aquatic life forms that occurred following the Miocene Climatic Optimum
(18 to 16 Ma), around 14.8 to 14.5 million years ago, during the Langhian stage
of the mid-Miocene. A major and permanent cooling step occurred between
14.8 and 14.1 Ma, associated with increased production of cold Antarctic deep
waters and a major growth of the East Antarctic ice sheet. A Middle Miocene
$\delta^{18}O$ increase, that is, a relative increase in the heavier isotope of oxygen, has
been noted in the Pacific, the Southern Ocean and the South Atlantic.

Further reading

- Cox, C. Barry & Moore, Peter D. (1993): *Biogeography. An ecological and evolutionary approach* (5th ed.). Blackwell Scientific Publications, Cambridge. ISBN 0-632-02967-6
- Ogg, Jim (2004): " Overview of Global Boundary Stratotype Sections and Points (GSSP's)[440]". Retrieved 2006-04-30.

External links

Wikimedia Commons has media related to *Miocene*.

Wikisource has original works on the topic: *Cenozoic#Neogene*

- PBS Deep Time: Miocene[441]
- UCMP Berkeley Miocene Epoch Page[442]
- Miocene Microfossils: 200+ images of Miocene Foraminifera[443]
- Human Timeline (Interactive)[444] – Smithsonian, National Museum of Natural History (August 2016).

Pliocene

System/ Period	Series/ Epoch	Stage/ Age	Age (Ma)	
Quaternary	Pleistocene	Gelasian	younger	
Neogene	Pliocene	Piacenzian	2.58	3.600
		Zanclean	3.600	5.333
	Miocene	Messinian	5.333	7.246
		Tortonian	7.246	11.63
		Serravallian	11.63	13.82
		Langhian	13.82	15.97
		Burdigalian	15.97	20.44
		Aquitanian	20.44	23.03
Paleogene	Oligocene	Chattian	older	

Subdivision of the Neogene Period
according to the ICS, as of 2017.

The **Pliocene** (/'plaɪəˌsiːn/;[445,446] also **Pleiocene**[447]) Epoch is the epoch in the geologic timescale that extends from 5.333 million to 2.58[448] million years BP. It is the second and youngest epoch of the Neogene Period in the Cenozoic Era. The Pliocene follows the Miocene Epoch and is followed by the Pleistocene Epoch. Prior to the 2009 revision of the geologic time scale, which placed the four most recent major glaciations entirely within the Pleistocene, the Pliocene also included the Gelasian stage, which lasted from 2.588 to 1.806 million years ago, and is now included in the Pleistocene.

As with other older geologic periods, the geological strata that define the start and end are well identified but the exact dates of the start and end of the epoch are slightly uncertain. The boundaries defining the Pliocene are not set at an easily identified worldwide event but rather at regional boundaries between the warmer Miocene and the relatively cooler Pliocene. The upper boundary was set at the start of the Pleistocene glaciations.

Etymology

Hominin timeline

Axis scale: million years

🖐

Also see: *Life timeline* and *Nature timeline*

Charles Lyell (later Sir Charles) gave the Pliocene its name in *Principles of Geology* (volume 3, 1833).[449]

The word *pliocene* comes from the Greek words πλεῖον (*pleion*, "more") and καινός (*kainos*, "new" or "recent") and means roughly "continuation of the recent", referring to the essentially modern marine mollusc fauna.

H.W. Fowler called the term *Pliocene* (like other geological jargon such as *pleistocene* and *miocene*) a "regrettable barbarism" and an indication that even "a good classical scholar" such as Lyell should have requested a philologist's help when coining words.

To summarize the usage of these *"regrettable barbarisms"* in the labelling of the Cenozoic ("recent life") era (from youngest to oldest):

Epoch	Lit-erally	First Element			Second Element		
		Greek	Translit-eration	Meaning	Greek	Transliteration	Mean-ing
Pleis-tocene	most-new	πλεῖστος	*pleîstos*	"most"	καινός	*kainós* (Latinized as *cænus*)	"new"
Pliocene	more-new	πλεῖον	*pleion*	"more"			
Miocene	less-new	μείων	*meiōn*	"less"			
Oligocene	few-new	ὀλίγος	*oligos*	"few"			
Eocene	dawn-new	ἠώς	*ēṓs*	"dawn"			
Paleocene	old-new	παλαιός	*palaios*	"old(er)"			

with the understanding that these are all *new* relative to the Mesozoic ("middle life" - the age of dinosaurs) and Paleozoic ("old life" - Trilobites, coal forests, and the earliest Synapsida) eras.

Subdivisions

In the official timescale of the ICS, the Pliocene is subdivided into two stages. From youngest to oldest they are:

- Piacenzian (3.600–2.58 Ma)
- Zanclean (5.333–3.600 Ma)

The Piacenzian is sometimes referred to as the Late Pliocene, whereas the Zanclean is referred to as the Early Pliocene.

In the system of

- North American Land Mammal Ages (NALMA) include Hemphillian (9–4.75 Ma), and Blancan (4.75–1.806 Ma). The Blancan extends forward into the Pleistocene.
- South American Land Mammal Ages (SALMA) include Montehermosan (6.8–4.0 Ma), Chapadmalalan (4.0–3.0 Ma) and Uquian (3.0–1.2 Ma).

In the Paratethys area (central Europe and parts of western Asia) the Pliocene contains the Dacian (roughly equal to the Zanclean) and Romanian (roughly equal to the Piacenzian and Gelasian together) stages. As usual in stratigraphy, there are many other regional and local subdivisions in use.

In Britain the Pliocene is divided into the following stages (old to young): Gedgravian, Waltonian, Pre-Ludhamian, Ludhamian, Thurnian, Bramertonian or Antian, Pre-Pastonian or Baventian, Pastonian and Beestonian. In the Netherlands the Pliocene is divided into these stages (old to young): Brunssumian C, Reuverian A, Reuverian B, Reuverian C, Praetiglian, Tiglian A, Tiglian B, Tiglian C1-4b, Tiglian C4c, Tiglian C5, Tiglian C6 and Eburonian. The exact correlations between these local stages and the ICS stages is still a matter of detail.

Climate

The global average temperature in the mid-Pliocene (3.3–3 mya) was 2–3 °C higher than today, carbon dioxide levels were the same as today, and global sea level was 25 m higher. The northern hemisphere ice sheet was ephemeral before the onset of extensive glaciation over Greenland that occurred in the late Pliocene around 3 Ma. The formation of an Arctic ice cap is signaled by an abrupt shift in oxygen isotope ratios and ice-rafted cobbles in the North Atlantic and North Pacific ocean beds.[450] Mid-latitude glaciation was probably underway before the end of the epoch. The global cooling that occurred during the Pliocene may have spurred on the disappearance of forests and the spread of grasslands and savannas.

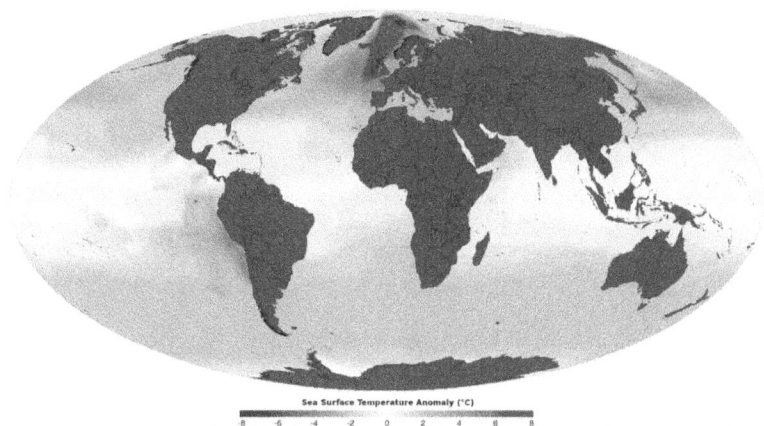

Figure 192: *Mid-Pliocene reconstructed annual sea surface temperature anomaly*

Figure 193: *19th century artist's impression of a Pliocene landscape*

Figure 194: *Examples of migrant species in the Americas after the formation of the Isthmus of Panama. Olive green silhouettes denote North American species with South American ancestors; blue silhouettes denote South American species of North American origin.*

Paleogeography

Continents continued to drift, moving from positions possibly as far as 250 km from their present locations to positions only 70 km from their current locations. South America became linked to North America through the Isthmus of Panama during the Pliocene, making possible the Great American Interchange and bringing a nearly complete end to South America's distinctive large marsupial predator and native ungulate faunas. The formation of the Isthmus had major consequences on global temperatures, since warm equatorial ocean currents were cut off and an Atlantic cooling cycle began, with cold Arctic and Antarctic waters dropping temperatures in the now-isolated Atlantic Ocean.

Africa's collision with Europe formed the Mediterranean Sea, cutting off the remnants of the Tethys Ocean. The border between the Miocene and the Pliocene is also the time of the Messinian salinity crisis.

Sea level changes exposed the land bridge between Alaska and Asia (Beringia).

Pliocene marine rocks are well exposed in the Mediterranean, India, and China. Elsewhere, they are exposed largely near shores.

Flora

The change to a cooler, dry, seasonal climate had considerable impacts on Pliocene vegetation, reducing tropical species worldwide. Deciduous forests proliferated, coniferous forests and tundra covered much of the north, and grasslands spread on all continents (except Antarctica). Tropical forests were limited to a tight band around the equator, and in addition to dry savannahs, deserts appeared in Asia and Africa.

Fauna

Both marine and continental faunas were essentially modern, although continental faunas were a bit more primitive than today. The first recognizable hominins, the australopithecines, appeared in the Pliocene.

The land mass collisions meant great migration and mixing of previously isolated species, such as in the Great American Interchange. Herbivores got bigger, as did specialized predators.

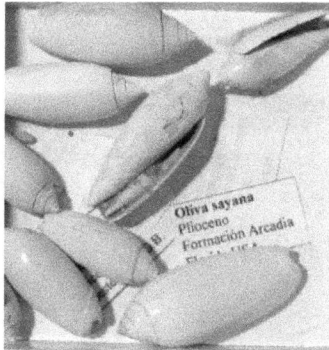

Figure 195: *The gastropod Oliva sayana, from the Pliocene of Florida.*

Figure 196: *The coral Cladocora from the Pliocene of Cyprus.*

Figure 197: *A gastropod and attached ser-pulid wormtube from the Pliocene of Cyprus.*

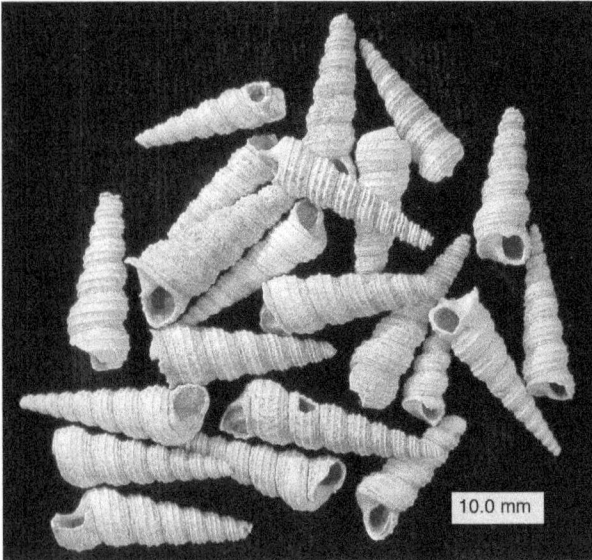

Figure 198: *The gastropod Turritella carinata from the Pliocene of Cyprus.*

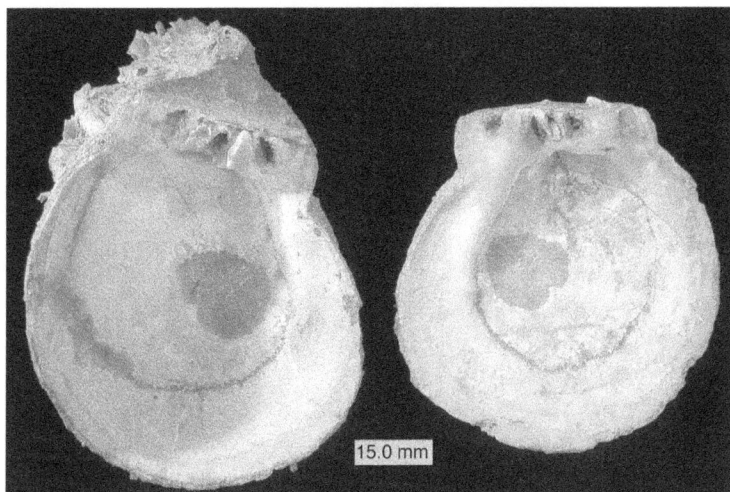

Figure 199: *The thorny oyster Spondylus right and left valve interiors from the Pliocene of Cyprus.*

Figure 200: *Articulated Spondylus from the Pliocene of Cyprus.*

Figure 201: *The limpet Diodora italica from the Pliocene of Cyprus.*

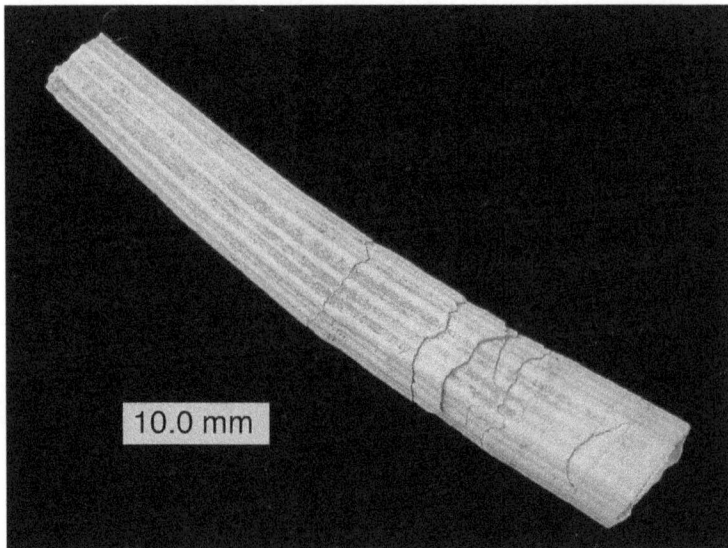

Figure 202: *The scaphopod Dentalium from the Pliocene of Cyprus.*

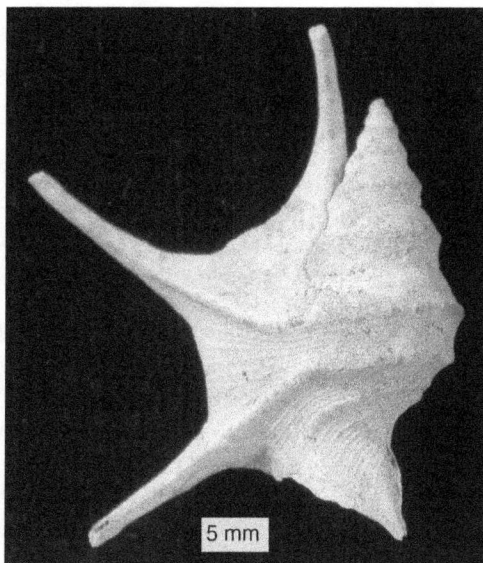

Figure 203: *The gastropod Aporrhais from the Pliocene of Cyprus.*

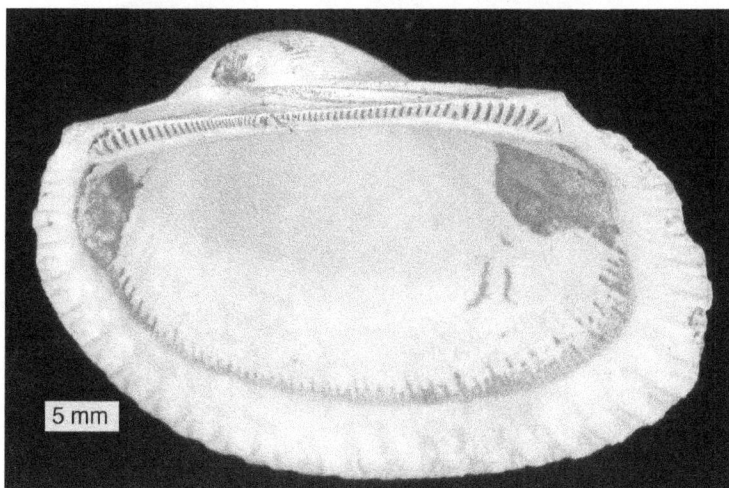

Figure 204: *The arcid bivalve Anadara from the Pliocene of Cyprus.*

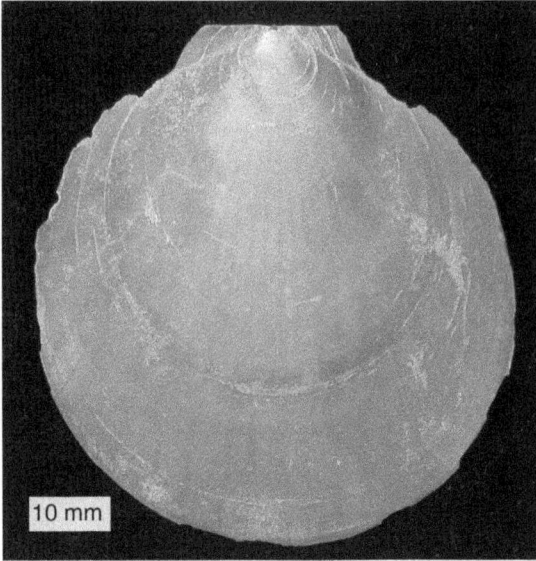

Figure 205: *The pectenid bivalve Ammu-
sium cristatum from the Pliocene of Cyprus.*

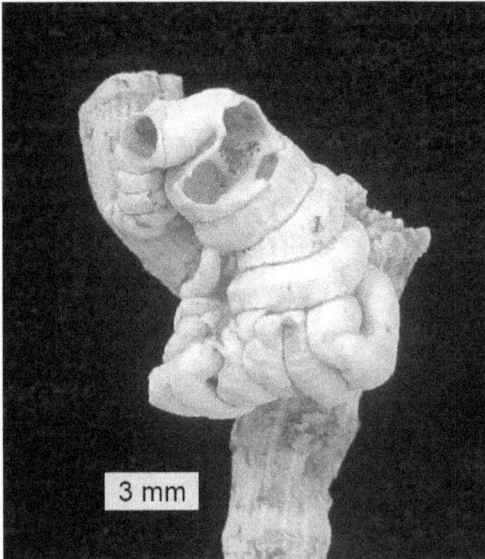

Figure 206: *Vermetid gastropod Petaloconchus intortus attached
to a branch of the coral Cladocora from the Pliocene of Cyprus.*

Mammals

In North America, rodents, large mastodons and gomphotheres, and opossums continued successfully, while hoofed animals (ungulates) declined, with camel, deer and horse all seeing populations recede. Rhinos, three toed horses (*Nannippus*), oreodonts, protoceratids, and chalicotheres became extinct. Borophagine dogs and *Agriotherium* became extinct, but other carnivores including the weasel family diversified, and dogs and short-faced bears did well. Ground sloths, huge glyptodonts, and armadillos came north with the formation of the Isthmus of Panama.

In Eurasia rodents did well, while primate distribution declined. Elephants, gomphotheres and stegodonts were successful in Asia, and hyraxes migrated north from Africa. Horse diversity declined, while tapirs and rhinos did fairly well. Cows and antelopes were successful, and some camel species crossed into Asia from North America. Hyenas and early saber-toothed cats appeared, joining other predators including dogs, bears and weasels.

Human evolution during the Pliocene

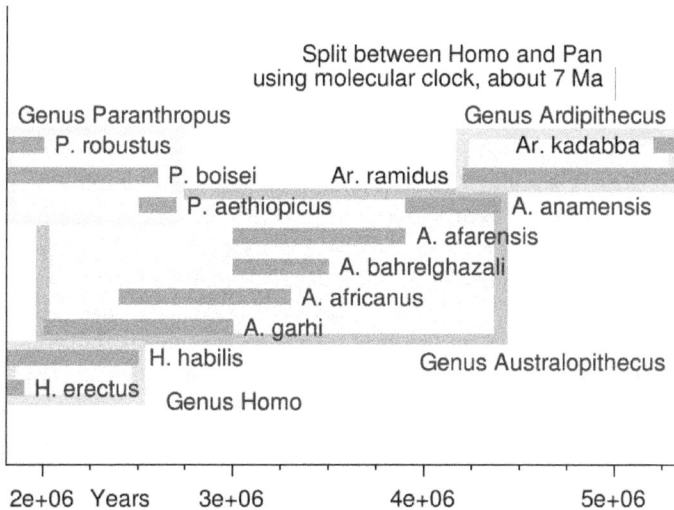

Split between Homo and Pan
using molecular clock, about 7 Ma

Genus Paranthropus   Genus Ardipithecus
P. robustus   Ar. kadabba
        P. boisei   Ar. ramidus
           P. aethiopicus   A. anamensis
                   A. afarensis
                   A. bahrelghazali
               A. africanus
           A. garhi
        H. habilis   Genus Australopithecus
H. erectus
           Genus Homo

2e+06 Years   3e+06   4e+06   5e+06

Africa was dominated by hoofed animals, and primates continued their evolution, with australopithecines (some of the first hominins) appearing in the late Pliocene. Rodents were successful, and elephant populations increased. Cows and antelopes continued diversification and overtaking pigs in numbers of species. Early giraffes appeared. Horses and modern rhinos came onto

Figure 207: *Pliocene mammals of North America*

the scene. Bears, dogs and weasels (originally from North America) joined cats, hyenas and civets as the African predators, forcing hyenas to adapt as specialized scavengers.

South America was invaded by North American species for the first time since the Cretaceous, with North American rodents and primates mixing with southern forms. Litopterns and the notoungulates, South American natives, were mostly wiped out, except for the macrauchenids and toxodonts, which managed to survive. Small weasel-like carnivorous mustelids, coatis and short-faced bears migrated from the north. Grazing glyptodonts, browsing giant ground sloths and smaller caviomorph rodents, pampatheres, and armadillos did the opposite, migrating to the north and thriving there.

The marsupials remained the dominant Australian mammals, with herbivore forms including wombats and kangaroos, and the huge *Diprotodon*. Carnivorous marsupials continued hunting in the Pliocene, including dasyurids, the dog-like thylacine and cat-like *Thylacoleo*. The first rodents arrived in Australia. The modern platypus, a monotreme, appeared.

Figure 208: *Titanis*

Birds

The predatory South American phorusrhacids were rare in this time; among the last was *Titanis*, a large phorusrhacid that migrated to North America and rivaled mammals as top predator. Other birds probably evolved at this time, some modern, some now extinct.

Reptiles and amphibians

Alligators and crocodiles died out in Europe as the climate cooled. Venomous snake genera continued to increase as more rodents and birds evolved. Rattlesnakes first appeared in the Pliocene. The modern species *Alligator mississippiensis*, having evolved in the Miocene, continued into the Pliocene, except with a more northern range; specimens have been found in very late Miocene deposits of Tennessee. Giant tortoises still thrived in North America, with genera like *Hesperotestudo*. Madtsoid snakes were still present in Australia. The amphibian order Allocaudata became extinct.

Oceans

Oceans continued to be relatively warm during the Pliocene, though they continued cooling. The Arctic ice cap formed, drying the climate and increasing cool shallow currents in the North Atlantic. Deep cold currents flowed from the Antarctic.

The formation of the Isthmus of Panama about 3.5 million years ago cut off the final remnant of what was once essentially a circum-equatorial current that had existed since the Cretaceous and the early Cenozoic. This may have contributed to further cooling of the oceans worldwide.

The Pliocene seas were alive with sea cows, seals, sea lions and sharks.

Supernovae

In 2002, Narciso Benítez et al. calculated that roughly 2 million years ago, around the end of the Pliocene epoch, a group of bright O and B stars called the Scorpius-Centaurus OB association passed within 130 light-years of Earth and that one or more supernova explosions gave rise to a feature known as the Local Bubble. Such a close explosion could have damaged the Earth's ozone layer and caused the extinction of some ocean life (at its peak, a supernova of this size could have the same absolute magnitude as an entire galaxy of 200 billion stars).[451]

Further reading

- Comins, Niel F.; William J. Kaufmann III (2005). *Discovering the Universe* (7th ed.). New York, NY: Susan Finnemore Brennan. ISBN 0-7167-7584-0.
- Gradstein, F.M.; Ogg, J.G. & Smith, A.G.; **2004**: *A Geologic Time Scale 2004*, Cambridge University Press.
- Ogg, Jim (June 2004). "Overview of Global Boundary Stratotype Sections and Points (GSSP's)"[452]. Archived from the original[453] on 2006-04-23. Retrieved 2006-04-30.
- Van Andel, Tjeerd H. (1994). *New Views on an Old Planet: a History of Global Change* (2nd ed.). Cambridge: Cambridge University Press. ISBN 0-521-44243-5.

External links

> Wikimedia Commons has media related to *Pliocene*.

> Wikisource has original works on the topic: *Cenozoic#Neogene*

- Mid-Pliocene Global Warming: NASA/GISS Climate Modeling[454]
- Palaeos Pliocene[455]
- PBS Change: Deep Time: Pliocene[456]
- Possible Pliocene supernova[457]
- "Supernova dealt deaths on Earth? Stellar blasts may have killed ancient marine life" *Science News Online*[458] retrieved February 2, 2002
- UCMP Berkeley Pliocene Epoch Page[459]
- Pliocene Microfossils: 100+ images of Pliocene Foraminifera[460]
- Human Timeline (Interactive)[461] – Smithsonian, National Museum of Natural History (August 2016).

Quaternary

Quaternary Period *2.58–0 million years ago* PreЄЄ OSD C P T J K P₃N	
Mean atmospheric O ₂ content over period duration	c. 20.8 vol %[462,463] (104 % of modern level
Mean atmospheric CO ₂ content over period duration	c. 250 ppm[464] (1 times pre-industrial level)
Mean surface temperature over period duration	c. 14 °C[465] (0 °C above modern level)

Quaternary (/kwəˈtɜːrnəri/) is the current and most recent of the three periods of the Cenozoic Era in the geologic time scale of the International Commission on Stratigraphy (ICS). It follows the Neogene Period and spans from 2.588 ± 0.005 million years ago to the present. The Quaternary Period is divided into two epochs: the Pleistocene (2.588 million years ago to 11.7 thousand years ago) and the Holocene (11.7 thousand years ago to today). The informal term "Late Quaternary" refers to the past 0.5–1.0 million years.[466]

The Quaternary Period is typically defined by the cyclic growth and decay of continental ice sheets associated with Milankovitch cycles and the associated climate and environmental changes that occurred.

Research history

In 1759 Giovanni Arduino proposed that the geological strata of northern Italy could be divided into four successive formations or "orders" (Italian: *quattro ordini*).[467] The term "quaternary" was introduced by Jules Desnoyers in 1829 for sediments of France's Seine Basin that seemed clearly to be younger than Tertiary Period rocks.[468]

The Quaternary Period follows the Neogene Period and extends to the present. The Quaternary covers the time span of glaciations classified as the Pleistocene, and includes the present interglacial time-period, the Holocene.

This places the start of the Quaternary at the onset of Northern Hemisphere glaciation approximately 2.6 million years ago. Prior to 2009, the Pleistocene was defined to be from 1.805 million years ago to the present, so the current definition of the Pleistocene includes a portion of what was, prior to 2009, defined as the Pliocene.

Subdivisions of the Quaternary System				
System/ Period	Series/ Epoch	Stage/ Age	Age (Ma)	
Quaternary	Holocene	Meghalayan	0	0.0042
		Northgrippian	0.0042	0.00833
		Greenlandian	0.00833	0.0117
	Pleistocene	'Tarantian'	0.0117	0.126
		'Chibanian'	0.126	0.781
		Calabrian	0.781	1.80
		Gelasian	1.80	2.58
Neogene	Pliocene	Piacenzian	older	

Subdivision of the Quaternary period according to the ICS, as of 2018.
Dates are relative to the year 2000 (eg. Greenlandian began 8,333 years before 2000), except for the
Meghalayan which began 4,200 years BP ie. before 1950.Wikipedia:Please clarify
'Chibanian' and 'Tarantian' are informal, unofficial names proposed to replace the also informal, unofficial
'Middle Pleistocene' and 'Upper Pleistocene' subseries/subepochs respectively.
In Europe and North America, the Holocene is subdivided into Preboreal, Boreal, Atlantic, Subboreal,
and Subatlantic stages of the Blytt–Sernander time scale. There are many regional subdivisions for the
Upper or Late Pleistocene; usually these represent locally recognized cold (glacial) and warm (interglacial)
periods. The last glacial period ends with the cold Younger Dryas substage.

Quaternary stratigraphers usually worked with regional subdivisions. From the 1970s, the International Commission on Stratigraphy (ICS) tried to make a single geologic time scale based on GSSP's, which could be used internationally. The Quaternary subdivisions were defined based on biostratigraphy instead of paleoclimate.

This led to the problem that the proposed base of the Pleistocene was at 1.805 Mya, long after the start of the major glaciations of the northern hemisphere. The ICS then proposed to abolish use of the name Quaternary altogether, which appeared unacceptable to the International Union for Quaternary Research (INQUA).

In 2009, it was decided to make the Quaternary the youngest period of the Cenozoic Era with its base at 2.588 Mya and including the Gelasian stage, which was formerly considered part of the Neogene Period and Pliocene Epoch.[469]

The Anthropocene has been proposed as a third epoch as a mark of the anthropogenic impact on the global environment starting with the Industrial Revolution, or about 200 years ago. The Anthropocene is not officially designated by the ICS, however, but a working group is currently aiming to complete a proposal for the creation of an epoch or sub-period by 2016.

Geology

The 2.6 million years of the Quaternary represents the time during which recognizable humans existed. Over this geologically short time period, there has been relatively little change in the distribution of the continents due to plate tectonics.

The Quaternary geological record is preserved in greater detail than that for earlier periods.

The major geographical changes during this time period included the emergence of the Strait of Bosphorus and Skagerrak during glacial epochs, which respectively turned the Black Sea and Baltic Sea into fresh water, followed by their flooding (and return to salt water) by rising sea level; the periodic filling of the English Channel, forming a land bridge between Britain and the European mainland; the periodic closing of the Bering Strait, forming the land bridge between Asia and North America; and the periodic flash flooding of Scablands of the American Northwest by glacial water.

The current extent of Hudson Bay, the Great Lakes and other major lakes of North America are a consequence of the Canadian Shield's readjustment since the last ice age; different shorelines have existed over the course of Quaternary time.

Climate

The climate was one of periodic glaciations with continental glaciers moving as far from the poles as 40 degrees latitude. There was a major extinction of large mammals in Northern areas at the end of the Pleistocene Epoch. Many forms such as saber-toothed cats, mammoths, mastodons, glyptodonts, etc., became extinct worldwide. Others, including horses, camels and American cheetahs became extinct in North America.

Quaternary glaciation

Glaciation took place repeatedly during the Quaternary Ice Age – a term coined by Schimper in 1839 that began with the start of the Quaternary about 2.58 Mya and continues to the present day.

Figure 209: *Artist's impression of Earth during the Last Glacial Maximum*

Last glacial period

In 1821, a Swiss engineer, Ignaz Venetz, presented an article in which he suggested the presence of traces of the passage of a glacier at a considerable distance from the Alps. This idea was initially disputed by another Swiss scientist, Louis Agassiz, but when he undertook to disprove it, he ended up affirming his colleague's hypothesis. A year later, Agassiz raised the hypothesis of a great glacial period that would have had long-reaching general effects. This idea gained him international fame and led to the establishment of the Glacial Theory.

In time, thanks to the refinement of geology, it has been demonstrated that there were several periods of glacial advance and retreat and that past temperatures on Earth were very different from today. In particular, the Milankovitch cycles of Milutin Milankovitch are based on the premise that variations in incoming solar radiation are a fundamental factor controlling Earth's climate.

During this time, substantial glaciers advanced and retreated over much of North America and Europe, parts of South America and Asia, and all of Antarctica. The Great Lakes formed and giant mammals thrived in parts of North America and Eurasia not covered in ice. These mammals became extinct when the glacial period Age ended about 11,700 years ago. Modern humans evolved about 315,000 years ago. During the Quaternary Period, mammals, flowering plants, and insects dominated the land. Wikipedia:Citation needed

Journals relating to the Quaternary Period

* *Boreas - An International Journal of Quaternary Research*
* *Geografiska Annaler* (only the title is in Swedish)
* *Journal of Quaternary Science*
* *Quaternary Geochronology*
* *Quaternary International*
* *Quaternary Research*
* *Quaternary Science Reviews*
* *The Quaternary Times*

External links

> Wikisource has original works on the topic: *Cenozoic#Quaternary*

* Subcommission on Quaternary Stratigraphy[470]
* Global correlation tables for the Quaternary[471]
* Gibbard, P.L., S. Boreham, K.M. Cohen and A. Moscariello, 2005, *Global chronostratigraphical correlation table for the last 2.7 million years v. 2005c.*[472], PDF version 220 KB. Subcommission on Quaternary Stratigraphy, Department of Geography, University of Cambridge, Cambridge, England
* Gibbard, P.L., S. Boreham, K.M. Cohen and A. Moscariello, 2007, *Global chronostratigraphical correlation table for the last 2.7 million years v. 2007b.*[473], jpg version 844 KB. Subcommission on Quaternary Stratigraphy, Department of Geography, University of Cambridge, Cambridge, England
* Silva, P.G. C. Zazo, T. Bardají, J. Baena, J. Lario, A. Rosas, J. Van der Made. 2009, « Tabla Cronoestratigráfíca del Cuaternario en la Península Ibérica - V.2[474] ». [Versión PDF, 3.6 Mb]. Asociación Española para el Estudio del Cuaternario (AEQUA), Departamento de Geología, Universidad de Salamanca, Spain. (Correlation chart of European Quaternary and cultural stages and fossils)
* Welcome to the XVIII INQUA-Congress, Bern, 2011[475]
* ⬗ Media related to Quaternary at Wikimedia Commons

Pleistocene

Subdivisions of the Quaternary System				
System/ Period	**Series/ Epoch**	**Stage/ Age**	**Age (Ma)**	
Quaternary	Holocene	Meghalayan	0	0.0042
		Northgrippian	0.0042	0.00833
		Greenlandian	0.00833	0.0117
	Pleistocene	'Tarantian'	0.0117	0.126
		'Chibanian'	0.126	0.781
		Calabrian	0.781	1.80
		Gelasian	1.80	2.58
Neogene	Pliocene	Piacenzian	**older**	

Subdivision of the Quaternary period according to the ICS, as of 2018.
Dates are relative to the year 2000 (eg. Greenlandian began 8,333 years before 2000), except for the Meghalayan which began 4,200 years BP ie. before 1950.Wikipedia:Please clarify
'Chibanian' and 'Tarantian' are informal, unofficial names proposed to replace the also informal, unofficial 'Middle Pleistocene' and 'Upper Pleistocene' subseries/subepochs respectively.
In Europe and North America, the Holocene is subdivided into Preboreal, Boreal, Atlantic, Subboreal, and Subatlantic stages of the Blytt–Sernander time scale. There are many regional subdivisions for the Upper or Late Pleistocene; usually these represent locally recognized cold (glacial) and warm (interglacial) periods. The last glacial period ends with the cold Younger Dryas substage.

The **Pleistocene** (/ˈplaɪstəˌsiːn, <wbr />-toʊ-/, often colloquially referred to as the **Ice Age**) is the geological epoch which lasted from about 2,588,000 to 11,700 years ago, spanning the world's most recent period of repeated glaciations. The end of the Pleistocene corresponds with the end of the last glacial period and also with the end of the Paleolithic age used in archaeology.

The Pleistocene is the first epoch of the Quaternary Period or sixth epoch of the Cenozoic Era.[476] In the ICS timescale, the Pleistocene is divided into four stages or ages, the Gelasian, Calabrian, Middle Pleistocene (unofficially the 'Chibanian') and Upper Pleistocene (unofficially the 'Tarantian').[477]</ref> In addition to this international subdivision, various regional subdivisions are often used.

Before a change finally confirmed in 2009 by the International Union of Geological Sciences, the time boundary between the Pleistocene and the preceding Pliocene was regarded as being at 1.806 million years Before Present (BP), as opposed to the currently accepted 2.588 million years BP: publications from the preceding years may use either definition of the period.

Etymology

Charles Lyell introduced the term "pleistocene" in 1839 to describe strata in Sicily that had at least 70% of their molluscan fauna still living today. This distinguished it from the older Pliocene Epoch, which Lyell had originally thought to be the youngest fossil rock layer. He constructed the name "Pleistocene" ("Most New" or "Newest") from the Greek πλεῖστος, *pleīstos*, "most", and καινός, *kainós* (latinized as *cænus*), "new";[478] this contrasting with the immediately preceding Pliocene ("More New" or "Newer", from πλείων, *pleíōn*, "more", and *kainós*; usual spelling: Pliocene), and the immediately subsequent Holocene ("wholly new" or "entirely new", from ὅλος, *hólos*, "whole", and *kainós*) epoch, which extends to the present time.

Dating

Hominin timeline

Axis scale: million years

✍

Also see: *Life timeline* and *Nature timeline*

The Pleistocene has been dated from 2.588 million (±0.005) to 11,700 years BP with the end date expressed in radiocarbon years as 10,000 carbon-14 years BP.[479] It covers most of the latest period of repeated glaciation, up to and including the Younger Dryas cold spell. The end of the Younger Dryas has been dated to about 9640 BC (11,654 calendar years BP). It was not until after the development of radiocarbon dating, however, that Pleistocene archaeological excavations shifted to stratified caves and rock-shelters as opposed to open-air river-terrace sites.

In 2009 the International Union of Geological Sciences (IUGS) confirmed a change in time period for the Pleistocene, changing the start date from 1.806 to 2.588 million years BP, and accepted the base of the Gelasian as the base of the Pleistocene, namely the base of the Monte San Nicola GSSP.[480] The IUGS has yet to approve a type section, Global Boundary Stratotype Section and Point (GSSP), for the upper Pleistocene/Holocene boundary (*i.e.* the upper boundary). The proposed section is the *North Greenland Ice Core Project* ice core 75° 06' N 42° 18' W. The lower boundary of the Pleistocene Series is formally defined magnetostratigraphically as the base of the Matuyama (C2r) chronozone, isotopic stage 103. Above this point there are notable extinctions of the calcareous nanofossils: *Discoaster pentaradiatus* and *Discoaster surculus*.[481]

The Pleistocene covers the recent period of repeated glaciations. The name Plio-Pleistocene has, in the past, been used to mean the last ice age. The revised definition of the Quaternary, by pushing back the start date of the Pleistocene to 2.58 Ma, results in the inclusion of all the recent repeated glaciations within the Pleistocene.

Paleogeography and climate

The modern continents were essentially at their present positions during the Pleistocene, the plates upon which they sit probably having moved no more than 100 km relative to each other since the beginning of the period.

According to Mark Lynas (through collected data), the Pleistocene's overall climate could be characterized as a continuous El Niño with trade winds in the south Pacific weakening or heading east, warm air rising near Peru, warm water spreading from the west Pacific and the Indian Ocean to the east Pacific, and other El Niño markers.[482]

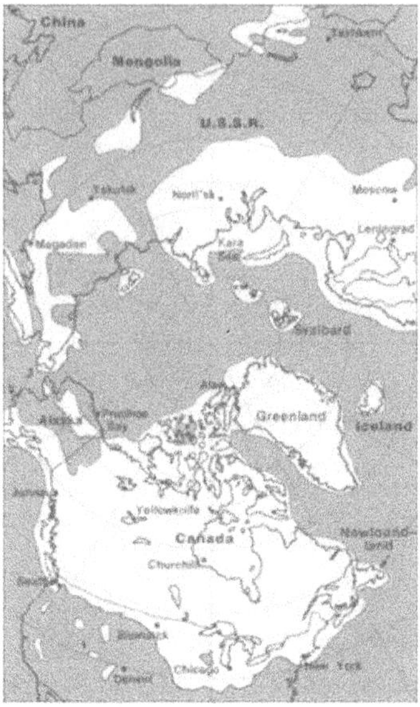

Figure 210: *The maximum extent of glacial ice in the north polar area during the Pleistocene period.*

Glacial features

Pleistocene climate was marked by repeated glacial cycles in which continental glaciers pushed to the 40th parallel in some places. It is estimated that, at maximum glacial extent, 30% of the Earth's surface was covered by ice. In addition, a zone of permafrost stretched southward from the edge of the glacial sheet, a few hundred kilometres in North America, and several hundred in Eurasia. The mean annual temperature at the edge of the ice was -6 °C (21 °F); at the edge of the permafrost, 0 °C (32 °F).

Each glacial advance tied up huge volumes of water in continental ice sheets 1,500 to 3,000 metres (4,900–9,800 ft) thick, resulting in temporary sea-level drops of 100 metres (300 ft) or more over the entire surface of the Earth. During interglacial times, such as at present, drowned coastlines were common, mitigated by isostatic or other emergent motion of some regions.

The effects of glaciation were global. Antarctica was ice-bound throughout the Pleistocene as well as the preceding Pliocene. The Andes were covered in the south by the Patagonian ice cap. There were glaciers in New Zealand and Tasmania. The current decaying glaciers of Mount Kenya, Mount Kilimanjaro, and the Ruwenzori Range in east and central Africa were larger. Glaciers existed in the mountains of Ethiopia and to the west in the Atlas mountains.

In the northern hemisphere, many glaciers fused into one. The Cordilleran ice sheet covered the North American northwest; the east was covered by the Laurentide. The Fenno-Scandian ice sheet rested on northern Europe, including Great Britain; the Alpine ice sheet on the Alps. Scattered domes stretched across Siberia and the Arctic shelf. The northern seas were ice-covered.

South of the ice sheets large lakes accumulated because outlets were blocked and the cooler air slowed evaporation. When the Laurentide ice sheet retreated, north central North America was totally covered by Lake Agassiz. Over a hundred basins, now dry or nearly so, were overflowing in the North American west. Lake Bonneville, for example, stood where Great Salt Lake now does. In Eurasia, large lakes developed as a result of the runoff from the glaciers. Rivers were larger, had a more copious flow, and were braided. African lakes were fuller, apparently from decreased evaporation. Deserts on the other hand were drier and more extensive. Rainfall was lower because of the decreases in oceanic and other evaporation.

It has been estimated that during the Pleistocene, the East Antarctic Ice Sheet thinned by at least 500 meters, and that thinning since the Last Glacial Maximum is less than 50 meters and probably started after ca 14 ka.

Major events

Over 11 major glacial events have been identified, as well as many minor glacial events. A major glacial event is a general glacial excursion, termed a "glacial." Glacials are separated by "interglacials". During a glacial, the glacier experiences minor advances and retreats. The minor excursion is a "stadial"; times between stadials are "interstadials".

These events are defined differently in different regions of the glacial range, which have their own glacial history depending on latitude, terrain and climate. There is a general correspondence between glacials in different regions. Investigators often interchange the names if the glacial geology of a region is in the process of being defined. However, it is generally incorrect to apply the name of a glacial in one region to another.

For most of the 20th century only a few regions had been studied and the names were relatively few. Today the geologists of different nations are taking more of an interest in Pleistocene glaciology. As a consequence, the number of names is expanding rapidly and will continue to expand. Many of the advances and stadials remain unnamed. Also, the terrestrial evidence for some of them has been erased or obscured by larger ones, but evidence remains from the study of cyclical climate changes.

The glacials in the following tables show *historical* usages, are a simplification of a much more complex cycle of variation in climate and terrain, and

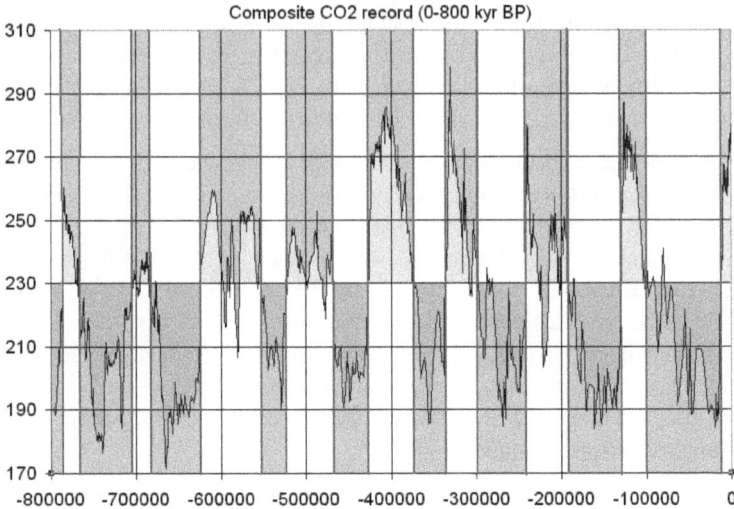

Figure 211: *Ice ages as reflected in atmospheric*
CO_2, stored in bubbles from glacial ice of Antarctica.

are generally no longer used. These names have been abandoned in favor of numeric data because many of the correlations were found to be either inexact or incorrect and more than four major glacials have been recognized since the historical terminology was established.[483,484]

Historical names of the "four major" glacials in four regions.

Region	Glacial 1	Glacial 2	Glacial 3	Glacial 4
Alps	Günz	Mindel	Riss	Würm
North Europe	Eburonian	Elsterian	Saalian	Weichselian
British Isles	Beestonian	Anglian	Wolstonian	Devensian
Midwest U.S.	Nebraskan	Kansan	Illinoian	Wisconsinan

Historical names of interglacials.

Region	Interglacial 1	Interglacial 2	Interglacial 3
Alps	Günz-Mindel	Mindel-Riss	Riss-Würm
North Europe	Waalian	Holsteinian	Eemian
British Isles	Cromerian	Hoxnian	Ipswichian
Midwest U.S.	Aftonian	Yarmouthian	Sangamonian

Corresponding to the terms glacial and interglacial, the terms pluvial and interpluvial are in use (Latin: *pluvia*, rain). A pluvial is a warmer period of increased rainfall; an interpluvial, of decreased rainfall. Formerly a pluvial was thought to correspond to a glacial in regions not iced, and in some cases it does. Rainfall is cyclical also. Pluvials and interpluvials are widespread.

There is no systematic correspondence of pluvials to glacials, however. Moreover, regional pluvials do not correspond to each other globally. For example, some have used the term "Riss pluvial" in Egyptian contexts. Any coincidence is an accident of regional factors. Only a few of the names for pluvials in restricted regions have been strategraphically defined.

Palaeocycles

The sum of transient factors acting at the Earth's surface is cyclical: climate, ocean currents and other movements, wind currents, temperature, etc. The waveform response comes from the underlying cyclical motions of the planet, which eventually drag all the transients into harmony with them. The repeated glaciations of the Pleistocene were caused by the same factors.

Milankovitch cycles

Glaciation in the Pleistocene was a series of glacials and interglacials, stadials and interstadials, mirroring periodic changes in climate. The main factor at work in climate cycling is now believed to be Milankovitch cycles. These are periodic variations in regional and planetary solar radiation reaching the Earth caused by several repeating changes in the Earth's motion.

Milankovitch cycles cannot be the sole factor responsible for the variations in climate since they explain neither the long term cooling trend over the Plio-Pleistocene, nor the millennial variations in the Greenland Ice Cores. Milankovitch pacing seems to best explain glaciation events with periodicity of 100,000, 40,000, and 20,000 years. Such a pattern seems to fit the information on climate change found in oxygen isotope cores.

Oxygen isotope ratio cycles

In oxygen isotope ratio analysis, variations in the ratio of ^{18}O to ^{16}O (two isotopes of oxygen) by mass (measured by a mass spectrometer) present in the calcite of oceanic core samples is used as a diagnostic of ancient ocean temperature change and therefore of climate change. Cold oceans are richer in ^{18}O, which is included in the tests of the microorganisms (foraminifera) contributing the calcite.

A more recent version of the sampling process makes use of modern glacial ice cores. Although less rich in ^{18}O than sea water, the snow that fell on the glacier year by year nevertheless contained ^{18}O and ^{16}O in a ratio that depended on the mean annual temperature.

Temperature and climate change are cyclical when plotted on a graph of temperature versus time. Temperature coordinates are given in the form of a deviation from today's annual mean temperature, taken as zero. This sort of graph is based on another of isotope ratio versus time. Ratios are converted to a percentage difference from the ratio found in standard mean ocean water (SMOW).

The graph in either form appears as a waveform with overtones. One half of a period is a Marine isotopic stage (MIS). It indicates a glacial (below zero) or an interglacial (above zero). Overtones are stadials or interstadials.

According to this evidence, Earth experienced 102 MIS stages beginning at about 2.588 Ma BP in the Early Pleistocene Gelasian. Early Pleistocene stages were shallow and frequent. The latest were the most intense and most widely spaced.

By convention, stages are numbered from the Holocene, which is MIS1. Glacials receive an even number; interglacials, odd. The first major glacial was MIS2-4 at about 85–11 ka BP. The largest glacials were 2, 6, 12, and 16; the warmest interglacials, 1, 5, 9 and 11. For matching of MIS numbers to named stages, see under the articles for those names.

Figure 212: *Pleistocene of Northern Spain showing woolly mammoth, cave lions eating a reindeer, tarpans, and woolly rhinoceros.*

Fauna

Both marine and continental faunas were essentially modern but with many more large land mammals such as Mammoths, Mastodons, *Diprotodon*, *Smilodon*, tiger, lion, Aurochs, Short-faced bear, giant sloths, *Gigantopithecus* and others. Isolated places such as Australia, Madagascar, New Zealand and islands in the Pacific saw the evolution of large birds and even reptiles such as the Elephant bird, moa, Haast's eagle, *Quinkana*, *Megalania* and *Meiolania*.

The severe climatic changes during the ice age had major impacts on the fauna and flora. With each advance of the ice, large areas of the continents became totally depopulated, and plants and animals retreating southwards in front of the advancing glacier faced tremendous stress. The most severe stress resulted from drastic climatic changes, reduced living space, and curtailed food supply. A major extinction event of large mammals (megafauna), which included mammoths, mastodons, saber-toothed cats, *glyptodons*, the woolly rhinoceros, various giraffids, such as the Sivatherium; ground sloths, Irish elk, cave bears, Gomphothere, dire wolves, and short-faced bears, began late in the Pleistocene and continued into the Holocene. Neanderthals also became extinct during this period. At the end of the last ice age, cold-blooded animals, smaller mammals like wood mice, migratory birds, and swifter animals like whitetail deer had replaced the megafauna and migrated north.

The extinctions hardly affected Africa but were especially severe in North America where native horses and camels were wiped out.

- Asian land mammal ages (ALMA) include Zhoukoudianian, Nihewanian, and Yushean.

Figure 213: *Pleistocene of South America showing Megatherium and two Glyptodon.*

- European land mammal ages (ELMA) include Gelasian (2.5—1.8 Ma).
- North American land mammal ages (NALMA) include Blancan (4.75–1.8), Irvingtonian (1.8–0.24) and Rancholabrean (0.24–0.01) in millions of years. The Blancan extends significantly back into the Pliocene.
- South American land mammal ages (SALMA) include Uquian (2.5–1.5), Ensenadan (1.5–0.3) and Lujanian (0.3–0.01) in millions of years. The Uquian previously extended significantly back into the Pliocene, although the new definition places it entirely within the Pleistocene.

In July 2018, a team of Russian scientists in collaboration with Princeton University announced that they had brought two female nematodes frozen in permafrost, from around 42,000 years ago, back to life. The two nematodes, at the time, were the oldest confirmed living animals on the planet.[485]

Humans

The evolution of anatomically modern humans took place during the Pleistocene. In the beginning of the Pleistocene *Paranthropus* species are still present, as well as early human ancestors, but during the lower Palaeolithic they disappeared, and the only hominin species found in fossilic records is *Homo erectus* for much of the Pleistocene. Acheulean lithics appear along with *Homo erectus*, some 1.8 million years ago, replacing the more primitive Oldowan industry used by *A. garhi* and by the earliest species of *Homo*. The Middle Paleolithic saw more varied speciation within *Homo*, including the appearance of *Homo sapiens* about 200,000 years ago.

According to mitochondrial timing techniques, modern humans migrated from Africa after the Riss glaciation in the Middle Palaeolithic during the Eemian Stage, spreading all over the ice-free world during the late Pleistocene.[486] A 2005 study posits that humans in this migration interbred with archaic human forms already outside of Africa by the late Pleistocene, incorporating archaic human genetic material into the modern human gene pool.

Hominin species during Pleistocene

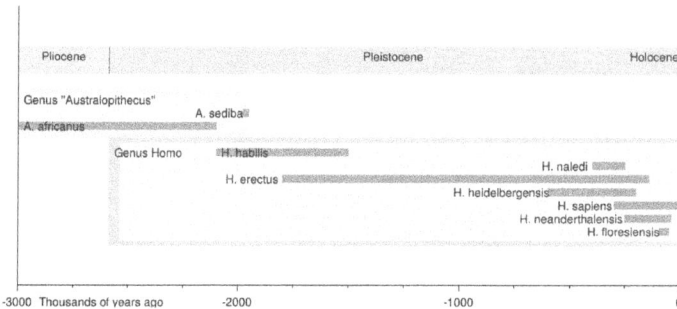

Deposits

Pleistocene non-marine sediments are found primarily in fluvial deposits, lakebeds, slope and loess deposits as well as in the large amounts of material moved about by glaciers. Less common are cave deposits, travertines and volcanic deposits (lavas, ashes). Pleistocene marine deposits are found primarily in shallow marine basins mostly (but with important exceptions) in areas within a few tens of kilometers of the modern shoreline. In a few geologically active areas such as the Southern California coast, Pleistocene marine deposits may be found at elevations of several hundred meters.

References

• Ogg, Jim; June, 2004, *Overview of Global Boundary Stratotype Sections and Points (GSSP's*, Stratigraphy.org[487], Accessed April 30, 2006.

External links

Wikimedia Commons has media related to *Pleistocene*.

Wikisource has original works on the topic: *Cenozoic#Quaternary*

- Late Pleistocene environments of the southern high plains[488], 1975, edited by Wendorf and Hester.
- Pleistocene Microfossils: 50+ images of Foraminifera[489]
- Stepanchuk V.N., Sapozhnykov I.V. Nature and man in the pleistocene of Ukraine. 2010[490]
- Human Timeline (Interactive)[491] – Smithsonian, National Museum of Natural History (August 2016).

Holocene

Subdivisions of the Quaternary System

System/ Period	Series/ Epoch	Stage/ Age	Age (Ma)	
Quaternary	Holocene	Meghalayan	0	0.0042
		Northgrippian	0.0042	0.00833
		Greenlandian	0.00833	0.0117
	Pleistocene	'Tarantian'	0.0117	0.126
		'Chibanian'	0.126	0.781
		Calabrian	0.781	1.80
		Gelasian	1.80	2.58
Neogene	Pliocene	Piacenzian	**older**	

Subdivision of the Quaternary period according to the ICS, as of 2018.
Dates are relative to the year 2000 (eg. Greenlandian began 8,333 years before 2000), except for the Meghalayan which began 4,200 years BP ie. before 1950.Wikipedia:Please clarify
'Chibanian' and 'Tarantian' are informal, unofficial names proposed to replace the also informal, unofficial 'Middle Pleistocene' and 'Upper Pleistocene' subseries/subepochs respectively.
In Europe and North America, the Holocene is subdivided into Preboreal, Boreal, Atlantic, Subboreal, and Subatlantic stages of the Blytt–Sernander time scale. There are many regional subdivisions for the Upper or Late Pleistocene; usually these represent locally recognized cold (glacial) and warm (interglacial) periods. The last glacial period ends with the cold Younger Dryas substage.

Holocene Epoch
This box:

- view
- talk
- edit[492]

↑ Pleistocene

Holocene

Preboreal (10.3–9 ka)
Boreal (9–7.5 ka)
Atlantic (7.5–5 ka)
Subboreal (5–2.5 ka)
Subatlantic (2.5 ka–present)

Look up *Holocene* in Wiktionary, the free dictionary.

The **Holocene** (/ˈhɒləˌsiːn, <wbr />ˈhoʊ-/) is the current geological epoch. It began approximately 11,650 cal years before present, after the last glacial period. The Holocene and the preceding Pleistocene together form the Quaternary period. The Holocene has been identified with the current warm period, known as MIS 1, and is considered by some to be an interglacial period.

The Holocene has seen the growth and impacts of the human species world-wide, including all its written history, development of major civilizations, and overall significant transition toward urban living in the present. Human impacts on modern-era Earth and its ecosystems may be considered of global significance for future evolution of living species, including approximately synchronous lithospheric evidence, or more recently hydrospheric and atmospheric evidence of human impacts. In July 2018, International Union of Geological Sciences split Holocene into three distinct subsections, Greenlandian (11,700 years ago to 8,326 years ago), Northgrippian (8,326 years ago to 4,200 years ago) and Meghalayan (4,200 years ago to the present), as proposed by International Commission on Stratigraphy. The boundary stratotype of Meghalayan is a speleothem in Mawmluh cave in India, and the global auxiliary stratotype is an ice core from Mount Logan in Canada.[493]

The name *Holocene* comes from the Ancient Greek words ὅλος (*holos*, whole or entire) and καινός (*kainos*, new), meaning "entirely recent".[494]

Overview

It is accepted by the International Commission on Stratigraphy that the Holocene started approximately 11,650 cal years BP. The Subcommission on Quaternary Stratigraphy quotes Gibbard and van Kolfschoten in Gradstein Ogg and Smith in stating the term 'Recent' as an alternative to Holocene is invalid and should not be used and also observe that the term Flandrian, derived from marine transgression sediments on the Flanders coast of Belgium has been used as a synonym for Holocene by authors who consider the last 10,000 years should have the same stage-status as previous interglacial events and thus be included in the Pleistocene. The International Commission on Stratigraphy, however, considers the Holocene an epoch following the Pleistocene and specifically the last glacial period. Local names for the last glacial period include the Wisconsinan in North America, the Weichselian in Europe, the Devensian in Britain, the Llanquihue in Chile and the Otiran in New Zealand.

The Holocene can be subdivided into five time intervals, or chronozones, based on climatic fluctuations:

- Preboreal (10 ka–9 ka),
- Boreal (9 ka–8 ka),
- Atlantic (8 ka–5 ka),
- Subboreal (5 ka–2.5 ka) and
- Subatlantic (2.5 ka–present).

Note: "ka" means "kilo-annum" i.e. 1,000 years (non-calibrated C14 dates)

The Blytt–Sernander classification of climatic periods initially defined by plant remains in peat mosses, is currently being explored. Geologists working in different regions are studying sea levels, peat bogs and ice core samples by a variety of methods, with a view toward further verifying and refining the Blytt–Sernander sequence. They find a general correspondence across Eurasia and North America, though the method was once thought to be of no interest. The scheme was defined for Northern Europe, but the climate changes were claimed to occur more widely. The periods of the scheme include a few of the final pre-Holocene oscillations of the last glacial period and then classify climates of more recent prehistory.Wikipedia:Citation needed

Paleontologists have not defined any faunal stages for the Holocene. If subdivision is necessary, periods of human technological development, such as the Mesolithic, Neolithic, and Bronze Age, are usually used. However, the time periods referenced by these terms vary with the emergence of those technologies in different parts of the world.Wikipedia:Citation needed

Climatically, the Holocene may be divided evenly into the Hypsithermal and Neoglacial periods; the boundary coincides with the start of the Bronze Age in Europe. According to some scholars, a third division, the Anthropocene, has now begun. The International Commission on Stratigraphy Subcommission on Quaternary Stratigraphy's working group on the 'Anthropocene' (a term coined by Paul Crutzen and Eugene Stoermer in 2000) note this term is used to denote the present time interval in which many geologically significant conditions and processes have been profoundly altered by human activities. The 'Anthropocene' is not a formally defined geological unit.

Geology

Continental motions due to plate tectonics are less than a kilometre over a span of only 10,000 years. However, ice melt caused world sea levels to rise about 35 m (115 ft) in the early part of the Holocene. In addition, many areas above about 40 degrees north latitude had been depressed by the weight of the Pleistocene glaciers and rose as much as 180 m (590 ft) due to post-glacial rebound over the late Pleistocene and Holocene, and are still rising today.

The sea level rise and temporary land depression allowed temporary marine incursions into areas that are now far from the sea. Holocene marine fossils are known, for example, from Vermont and Michigan. Other than higher-latitude temporary marine incursions associated with glacial depression, Holocene fossils are found primarily in lakebed, floodplain, and cave deposits. Holocene

Figure 214: *Holocene cinder cone volcano on State Highway 18 near Veyo, Utah*

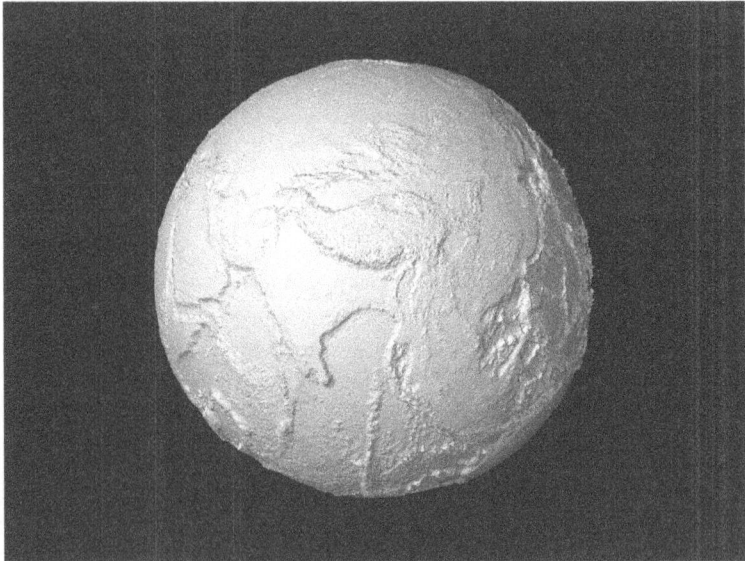

Figure 215: *Current Earth - without water (click/enlarge to "spin" 3D-globe).*

Holocene Temperature Variations

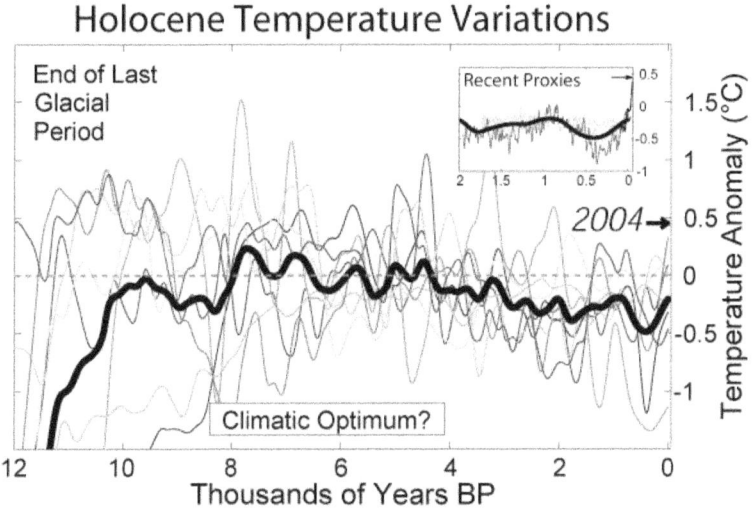

Figure 216: *Temperature variations during the Holocene*

marine deposits along low-latitude coastlines are rare because the rise in sea levels during the period exceeds any likely tectonic uplift of non-glacial origin.Wikipedia:Citation needed

Post-glacial rebound in the Scandinavia region resulted in the formation of the Baltic Sea. The region continues to rise, still causing weak earthquakes across Northern Europe. The equivalent event in North America was the rebound of Hudson Bay, as it shrank from its larger, immediate post-glacial Tyrrell Sea phase, to near its present boundaries.Wikipedia:Citation needed

Climate

Climate has been fairly stable over the Holocene. Ice core records show that before the Holocene there was global warming after the end of the last ice age and cooling periods, but climate changes became more regional at the start of the Younger Dryas. During the transition from the last glacial to the Holocene, the Huelmo–Mascardi Cold Reversal in the Southern Hemisphere began before the Younger Dryas, and the maximum warmth flowed south to north from 11,000 to 7,000 years ago. It appears that this was influenced by the residual glacial ice remaining in the Northern Hemisphere until the later date.Wikipedia:Citation needed

The Holocene climatic optimum (HCO) was a period of warming in which the global climate became warmer. However, the warming was probably not

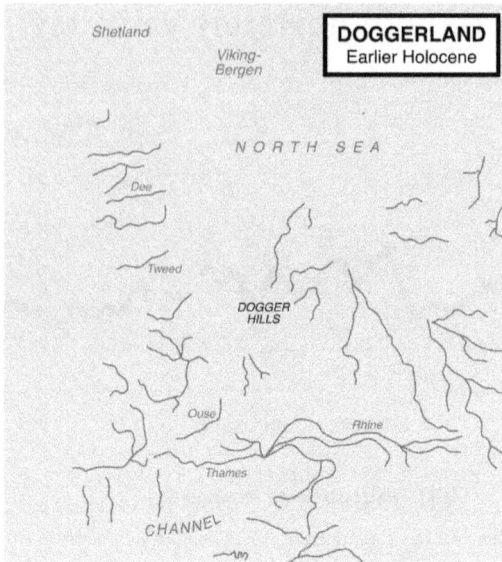

Figure 217: *Paleogeographic reconstruction of the North Sea approximately 9,000 years ago during the early Holocene and after the end of the last ice age.*

uniform across the world. This period of warmth ended about 5,500 years ago with the descent into the Neoglacial and concomitant Neopluvial. At that time, the climate was not unlike today's, but there was a slightly warmer period from the 10th–14th centuries known as the Medieval Warm Period. This was followed by the Little Ice Age, from the 13th or 14th century to the mid-19th century, which was a period of cooling.Wikipedia:Citation needed

Compared to glacial conditions, habitable zones have expanded northwards, reaching their northernmost point during the HCO. Greater moisture in the polar regions has caused the disappearance of steppe-tundra.Wikipedia:Citation needed

The temporal and spatial extent of Holocene climate change is an area of considerable uncertainty, with radiative forcing recently proposed to be the origin of cycles identified in the North Atlantic region. Climate cyclicity through the Holocene (Bond events) has been observed in or near marine settings and is strongly controlled by glacial input to the North Atlantic. Periodicities of ≈ 2500, ≈ 1500, and ≈ 1000 years are generally observed in the North Atlantic. At the same time spectral analyses of the continental record, which is remote from oceanic influence, reveal persistent periodicities of 1,000 and 500 years that may correspond to solar activity variations during the Holocene epoch. A 1,500-year cycle corresponding to the North Atlantic oceanic circulation may have had widespread global distribution in the Late Holocene.

Ecological developments

Animal and plant life have not evolved much during the relatively short Holocene, but there have been major shifts in the distributions of plants and animals. A number of large animals including mammoths and mastodons, saber-toothed cats like *Smilodon* and *Homotherium*, and giant sloths disappeared in the late Pleistocene and early Holocene—especially in North America, where animals that survived elsewhere (including horses and camels) became extinct. This extinction of American megafauna has been explained as caused by the arrival of the ancestors of Amerindians; though most scientists assert that climatic change also contributed. In addition, a controversial bolide impact over North America has been hypothesized to have triggered the Younger Dryas.

Throughout the world, ecosystems in cooler climates that were previously regional have been isolated in higher altitude ecological "islands".

The *8.2 ka event*, an abrupt cold spell recorded as a negative excursion in the $\delta^{18}O$ record lasting 400 years, is the most prominent climatic event occurring in the Holocene epoch, and may have marked a resurgence of ice cover. It has been suggested that this event was caused by the final drainage of Lake Agassiz, which had been confined by the glaciers, disrupting the thermohaline circulation of the Atlantic. Subsequent research, however, suggested that the discharge was probably superimposed upon a longer episode of cooler climate lasting up to 600 years and observed that the extent of the area affected was unclear.

Human developments

The beginning of the Holocene corresponds with the beginning of the Mesolithic age in most of Europe, but in regions such as the Middle East and Anatolia with a very early neolithisation, Epipaleolithic is preferred in place of Mesolithic. Cultures in this period include Hamburgian, Federmesser, and the Natufian culture, during which the oldest inhabited places still existing on Earth were first settled, such as Jericho in the Middle East. There is also evolving archeological evidence of proto-religion at locations such as Göbekli Tepe, as long ago as the 9th millennium BCE.

Both are followed by the aceramic Neolithic (Pre-Pottery Neolithic A and Pre-Pottery Neolithic B) and the pottery Neolithic. The Late Holocene brought advancements such as the bow and arrow and saw new methods of warfare in North America. Spear throwers and their large points were replaced by the bow

Figure 218: *Bronze bead necklace, Muséum de Toulouse*

and arrow with its small narrow points beginning in Oregon and Washington. Villages built on defensive bluffs indicate increased warfare, leading to food gathering in communal groups for protection rather than individual hunting. In Mesoamerica, transformations of natural environments have been a common feature at least since the mid-Holocene, mostly through the exploitation of wild plants and the establishment of crops.

Further reading

* Roberts, Neil (2014). *The Holocene: an environmental history* (3rd ed.). Malden, MA: Wiley-Blackwell. ISBN 978-1-4051-5521-2.
* Mackay, A. W.; Battarbee, R. W.; Birks, H. J. B.; et al., eds. (2003). *Global change in the Holocene*. London: Arnold. ISBN 0-340-76223-3.
* Hunt CO and Rabett RJ. (2013). Holocene landscape intervention and plant food production strategies in island and mainland Southeast Asia. Journal of Archaeological Science

External links

Wikimedia Commons has media related to *Holocene*.

Wikisource has original works on the topic: *Cenozoic#Quaternary*

Wikisourcehas the text of the 1911 *Encyclopædia Britannica*article *Holocene*.

- The Holocene epoch explained by the BBC[495]
- ghK Classification[496]

Appendix

References

[1] http://tools.wmflabs.org/timescale/?Ma=750
[2] http://tools.wmflabs.org/timescale/?Ma=600–540
[3] http://tools.wmflabs.org/timescale/?Ma=200
[4] http://tools.wmflabs.org/timescale/?Ma=40
[5] http://tools.wmflabs.org/timescale/?Ma=4,500
[6] http://tools.wmflabs.org/timescale/?Ma=4,400
[7] http://tools.wmflabs.org/timescale/?Ma=4100–3800
[8] http://tools.wmflabs.org/timescale/?Ma=4000–2500
[9] http://tools.wmflabs.org/timescale/?Ma=2500–541
[10] International Stratigraphic Chart 2008, International Commission on Stratigraphy https://www.webcitation.org/5nDdmUWXk?url=http://www.stratigraphy.org/upload/ISChart2008.pdf
[11] http://tools.wmflabs.org/timescale/?Ma=541–252
[12] http://tools.wmflabs.org/timescale/?Ma=447–444
[13] http://tools.wmflabs.org/timescale/?Ma=249
[14] http://tools.wmflabs.org/timescale/?Ma=290
[15] http://tools.wmflabs.org/timescale/?Ma=252–66
[16] Monroe and Wicander, 607.
[17] http://tools.wmflabs.org/timescale/?Ma=145.0
[18] http://tools.wmflabs.org/timescale/?Ma=66
[19] Dougal Dixon et al., *Atlas of Life on Earth*, (New York: Barnes & Noble Books, 2001), p. 215.
[20] http://tools.wmflabs.org/timescale/?Ma=56.0
[21] Hooker, J.J., "Tertiary to Present: Paleocene", pp. 459-465, Vol. 5. of Selley, Richard C., L. Robin McCocks, and Ian R. Plimer, Encyclopedia of Geology, Oxford: Elsevier Limited, 2005.
[22] http://tools.wmflabs.org/timescale/?Ma=33.9
[23] http://tools.wmflabs.org/timescale/?Ma=34
[24] http://tools.wmflabs.org/timescale/?Ma=23
[25] http://tools.wmflabs.org/timescale/?Ma=5.333
[26] http://tools.wmflabs.org/timescale/?Ma=2.588
[27] http://tools.wmflabs.org/timescale/?Ma=2.58
[28] https://www.webcitation.org/5QVjwZCzJ?url=http://www.tufts.edu/as/wright_center/cosmic_evolution/docs/splash.html
[29] http://www.sciam.com/article.cfm?chanID=sa006&colID=1&articleID=0005FA5D-5F7C-1333-9F7C83414B7F0000
[30] https://www.theguardian.com/science/story/0,3605,1671164,00.html
[31] http://www.johnkyrk.com/evolution.html
[32] https://web.archive.org/web/20030729055405/http://www.uwmc.uwc.edu/geography/Hutton/Hutton.htm
[33] http://issuu.com/sergioluisdasilva/docs/paleomaps_mollweide_longitude_0
[34] http://issuu.com/sergioluisdasilva/docs/paleomaps_mollweide_longitude_180
[35] http://www.bbc.co.uk/programmes/p005493g
[36] //en.wikipedia.org/w/index.php?title=Precambrian&action=edit
[37] Scientists discover 'oldest footprints on Earth' in southern China dating back 550 million years https://www.independent.co.uk/news/science/archaeology/oldest-fossil-footprints-china-found-discovered-yangtze-virginia-tech-a8386911.html *The Independent*
[38] Geological Society of America's "2009 GSA Geologic Time Scale." http://www.geosociety.org/science/timescale/
[39] also available at Stratigraphy.org: Precambrian subcommission http://www.stratigraphy.org/bak/precambrian/Bleeker_2004.pdf
[40] http://www.geology.wisc.edu/~valley/zircons/zircon_home.html
[41] //doi.org/10.1038/35051550

[42] //www.ncbi.nlm.nih.gov/pubmed/11196637
[43] http://adsabs.harvard.edu/abs/2004AuJES..51...31W
[44] //doi.org/10.1046/j.1400-0952.2003.01042.x
[45] http://www.scotese.com/precambr.htm
[46] //en.wikipedia.org/w/index.php?title=Hadean&action=edit
[47] http://tools.wmflabs.org/timescale/?Ma=4,533
[48] http://tools.wmflabs.org/timescale/?Ma=3,920
[49] http://tools.wmflabs.org/timescale/?Ma=3,850
[50] Early edition, published online before print.
[51] Solar System Exploration: Science & Technology: Science Features: View Feature http:
//solarsystem.nasa.gov/scitech/display.cfm?ST_ID=446
[52] ANU - Research School of Earth Sciences - ANU College of Science - Harrison http:
//wwwrses.anu.edu.au/admin/index.php?p=harrison
[53] ANU - OVC - MEDIA - MEDIA RELEASES - 2005 - NOVEMBER - 181105HARRISON-
CONTINENTS http://info.anu.edu.au/mac/Media/Media_Releases/_2005/_November/
_181105harrisoncontinents.asp
[54] A Cool Early Earth http://www.geology.wisc.edu/%7Evalley/zircons/cool_early/cool_early_
home.html
[55] http://adsabs.harvard.edu/abs/2008Natur.456..493H
[56] //doi.org/10.1038/nature07465
[57] //www.ncbi.nlm.nih.gov/pubmed/19037314
[58] http://www.geology.wisc.edu/~valley/zircons/zircon_home.html
[59] //doi.org/10.1038/35051550
[60] //www.ncbi.nlm.nih.gov/pubmed/11196637
[61] http://adsabs.harvard.edu/abs/2004AuJES..51...31W
[62] //doi.org/10.1046/j.1400-0952.2003.01042.x
[63] http://adsabs.harvard.edu/abs/2014E&PSL.405...85C
[64] //doi.org/10.1016/j.epsl.2014.08.015
[65] http://www.palaeos.org/Hadean
[66] https://web.archive.org/web/20071103114128/http://www.peripatus.gen.nz/paleontology/
Hadean.html
[67] http://www.astronoo.com/en/articles/hadean.html
[68] http://tools.wmflabs.org/timescale/?Ma=4000–2500
[69] The name "Archean" was coined by American geologist James Dwight Dana (1813–1895):
From p. 253: The Pre-Cambrian eon had been believed to be without life (azoic); however,
because fossils had been found in deposits that had been judged to belong to the Azoic age, " ...
I propose to use for the Azoic era and its rocks the general term *Archæn* (or Arche'an), from the
Greek ἀρχαιος, *pertaining to the beginning*.* "
[70] Early edition, published online before print.
[71] http://www.stratigraphy.org/bak/geowhen/stages/Archean.html
[72] http://www.utdallas.edu/~rjstern/PlateTectonicsStart/
[73] "Proterozoic – definition of Proterozoic in English from the Oxford dictionary" https://www.
oxforddictionaries.com/definition/english/Proterozoic. OxfordDictionaries.com. Retrieved
2016-01-20.
[74] "Proterozoic" https://www.merriam-webster.com/dictionary/Proterozoic. *Merriam-Webster
Dictionary*.
[75] Kearey, P., Klepeis, K., Vine, F., Precambrian Tectonics and the Supercontinent Cycle, Global
Tectonics, Third Edition, pp. 361–377, 2008.
[76] Mengel, F., Proterozoic History, Earth System: History and Variablility, volume 2, 1998.
[77] Huntly, C., The Mozambique Belt, Eastern Africa: Tectonic Evolution of the Mozambique
Ocean and Gondwana Amalgamation. The Geological Society of America. 2002.
[78] https://web.archive.org/web/20070702223746/http://www.palaeos.com/Proterozoic/
Proterozoic.htm
[79] There are several ways of pronouncing *Phanerozoic*, including /ˌfænərəˈzoʊɪk, <wbr />ˌfænrə-,
<wbr />-roʊ-/.<ref group=""> "Phanerozoic" https://en.oxforddictionaries.com/definition/
Phanerozoic. *Oxford Dictionaries*. Oxford University Press. Retrieved 2016-01-20.

[80] "Phanerozoic" https://www.merriam-webster.com/dictionary/Phanerozoic. *Merriam-Webster Dictionary*.

[81] Cohen, K.M.; Finney, S.C.; Gibbard, P.L.; Fan, J.-X. (2013). "International Chronostratigraphic Chart v 2015/01" http://www.stratigraphy.org/ICSchart/ChronostratChart2015-01.pdf (PDF). Episodes 36: 199-204. International Commission on Stratigraphy. Retrieved 2015-11-26.

[82] http://tools.wmflabs.org/timescale/?Ma=541–485

[83] See, e. g.,

[84] https://www.academia.edu/24472865/Phanerozoic_marine_biodiversity_follows_a_hyperbolic_trend

[85] //doi.org/10.1016/j.palwor.2007.01.002

[86] http://academiccommons.columbia.edu/download/fedora_content/download/ac:164668/CONTENT/Miller.Science.310.1293.pdf

[87] http://adsabs.harvard.edu/abs/2005Sci...310.1293M

[88] //doi.org/10.1126/science.1116412

[89] //www.ncbi.nlm.nih.gov/pubmed/16311326

[90] "Paleozoic" http://www.dictionary.com/browse/Paleozoic. *Dictionary.com Unabridged*. Random House.

[91] "Paleozoic" https://www.merriam-webster.com/dictionary/Paleozoic. *Merriam-Webster Dictionary*.

[92] The term "Palaeozoic" was coined by the British geologist Adam Sedgwick (1785–1873) in: ; see p. 685.

[93] http://tools.wmflabs.org/timescale/?Ma=541–251.902

[94] http://www.economist.com/node/16524904 The Economist

[95] http://tools.wmflabs.org/timescale/?Ma=510

[96] http://www.ucmp.berkeley.edu/paleozoic/paleozoic.php

[97] http://tools.wmflabs.org/timescale/?Ma=30

[98] http://tools.wmflabs.org/timescale/?Ma=445

[99] http://tools.wmflabs.org/timescale/?Ma=390

[100] http://www.stratigraphy.org/

[101] http://www.foraminifera.eu/querydb.php?&period=Carboniferous&aktion=suche

[102] Image:Sauerstoffgehalt-1000mj.svg

[103] File:OxygenLevel-1000ma.svg

[104] Image:Phanerozoic Carbon Dioxide.png

[105] Image:All palaeotemps.png

[106] Sedgwick and R. I. Murchison (1835) "On the Silurian and Cambrian systems, exhibiting the order in which the older sedimentary strata succeed each other in England and Wales," https://archive.org/stream/reportannualmee03meetgoog#page/n398/mode/1up/ Notices and Abstracts of Communications to the British Association for the Advancement of Science at the Dublin meeting, August 1835, pp. 59-61, in: *Report of the Fifth Meeting of the British Association for the Advancement of Science; held in Dublin in 1835* (1836). From p. 60: "Professor Sedgwick then described in descending order the groups of slate rocks, as they are seen in Wales and Cumberland. To the highest he gave the name of *Upper Cambrian* group. .. To the next inferior group he gave the name of *Middle Cambrian*. .. The *Lower Cambrian* group occupies the S.W. coast of Cærnarvonshire,"

[107] Schieber, 2007, pp. 53–71.

[108] A. Knoll, M. Walter, G. Narbonne, and N. Christie-Blick (2004) " The Ediacaran Period: A New Addition to the Geologic Time Scale. http://www.stratigraphy.org/bak/ediacaran/Knoll_et_al_2004a.pdf" Submitted on Behalf of the Terminal Proterozoic Subcommission of the International Commission on Stratigraphy.

[109] M.A. Fedonkin, B.S. Sokolov, M.A. Semikhatov, N.M.Chumakov (2007). " Vendian versus Ediacaran: priorities, contents, prospectives. http://vendian.net76.net/Vendian_vs_Ediacaran.htm " In: edited by M. A. Semikhatov " The Rise and Fall of the Vendian (Ediacaran) Biota. Origin of the Modern Biosphere. Transactions of the International Conference on the IGCP Project 493, August 20–31, 2007, Moscow. http://www.geosci.monash.edu.au/precsite/docs/workshop/moscow07/transaction.pdf" Moscow: GEOS.

[110] A. Ragozina, D. Dorjnamjaa, A. Krayushkin, E. Serezhnikova (2008). " *Treptichnus pedum* and the Vendian-Cambrian boundary http://vendian.net76.net/Treptichnus_pedum.htm". 33 Intern. Geol. Congr. 6–14 August 2008, Oslo, Norway. Abstracts. Section HPF 07 Rise and fall of the Ediacaran (Vendian) biota. P. 183.

[111] http://tools.wmflabs.org/timescale/?Ma=541

[112] http://tools.wmflabs.org/timescale/?Ma=509

[113] http://tools.wmflabs.org/timescale/?Ma=497

[114] http://tools.wmflabs.org/timescale/?Ma=485.4

[115] http://tools.wmflabs.org/timescale/?Ma=570

[116] Schieber et al., 2007, pp. 53–71.

[117] The Ordovician: Life's second big bang http://www.nuffieldfoundation.org/sites/default/files/the-ordovician-explosion-651.doc

[118] Collette & Hagadorn, 2010.

[119] Collette, Gass & Hagadorn, 2012

[120] Yochelson & Fedonkin, 1993.

[121] Getty & Hagadorn, 2008.

[122] Unicode Character 'LATIN CAPITAL LETTER C WITH BAR' (U+A792) https://www.fileformat.info/info/unicode/char/A792/index.htm. fileformat.info. Accessed 15 Jun 2015

[123] http://adsabs.harvard.edu/abs/2003Geo....31..431A

[124] //doi.org/10.1130/0091-7613%282003%29031%3C0431%3AEOCANA%3E2.0.CO%3B2

[125] //doi.org/10.1666/11-056.1

[126] //doi.org/10.1666/09-075.1

[127] //doi.org/10.1666/08-004.1

[128] //doi.org/10.1144/GSL.JGS.1852.008.01-02.26

[129] http://www.geol.umd.edu/~hcui/Reference/GeolTimeScale2012/Ch19-Cambrian.pdf

[130] http://www.sil.si.edu/smithsoniancontributions/Paleobiology/sc_RecordSingle.cfm?filename=SCtP-0074

[131] //doi.org/10.5479/si.00810266.74.1

[132] http://www.bbc.co.uk/programmes/p003k9bg

[133] http://www.trilobites.info/biostratigraphy.htm

[134] http://www.trilobite.info

[135] http://www.geo-lieven.com/erdzeitalter/kambrium/kambrium.htm

[136] http://www.scotese.com/

[137] https://web.archive.org/web/20030829130602/http://www.earth-pages.com/archive/geobiology.asp

[138] http://www.astrobio.net/news/article251.html

[139] Image:Sauerstoffgehalt-1000mj.svg

[140] File:OxygenLevel-1000ma.svg

[141] Image:Phanerozoic Carbon Dioxide.png

[142] Image:All palaeotemps.png

[143] Charles Lapworth (1879) "On the Tripartite Classification of the Lower Palaeozoic Rocks," https://books.google.com/books?id=JJpZAAAYAAJ&pg=PA1#v=onepage&q&f=false *Geological Magazine*, new series, **6** : 1-15. From pp. 13-14: "North Wales itself — at all events the whole of the great Bala district where Sedgwick first worked out the physical succession among the rocks of the intermediate or so-called *Upper Cambrian* or *Lower Silurian* system; and in all probability much of the Shelve and the Caradoc area, whence Murchison first published its distinctive fossils — lay within the territory of the Ordovices; ... Here, then, have we the hint for the appropriate title for the central system of the Lower Palaeozoics. It should be called the Ordovician System, after this old British tribe."

[144] Details on the Dapingian are available at

[145] Ordovician stratigraphy http://ordovician.stratigraphy.org/uploads/OrdChartHigh.jpg

[146] Explosion in marine biodiversity explained by climate change http://www2.cnrs.fr/en/1279.htm

[147] Humble moss helped to cool Earth and spurred on life https://www.bbc.co.uk/news/science-environment-16814669

[148] Palaeos Paleozoic : Ordovician : The Ordovician Period http://www.palaeos.com/Paleozoic/Ordovician/Ordovician.htm

[149] A Guide to the Orders of Trilobites http://www.trilobites.info/

[150] Image:Sauerstoffgehalt-1000mj.svg

[151] File:OxygenLevel-1000ma.svg

[152] Jeff Hecht, High-carbon ice age mystery solved https://www.newscientist.com/article/dn18618-highcarbon-ice-age-mystery-solved.html, *New Scientist*, 8 March 2010 (retrieved 30 June 2014)

[153] Emiliani, Cesare. (1992). *Planet Earth : Cosmology, Geology, & the Evolution of Life & the Environment* (Cambridge University Press) p. 491

[154] https://onlinelibrary.wiley.com/doi/abs/10.1111/pala.12188

[155] https://web.archive.org/web/20060423084018/http://www.stratigraphy.org/gssp.htm

[156] http://www.stratigraphy.org/gssp.htm

[157] http://www.anr.state.vt.us/dec/geo/chazytxt.htm

[158] http://members.wri.com/jeffb/Fossils

[159] http://www.drydredgers.org

[160] Image:Sauerstoffgehalt-1000mj.svg

[161] File:OxygenLevel-1000ma.svg

[162] Image:Phanerozoic Carbon Dioxide.png

[163] Image:All palaeotemps.png

[164] http://tools.wmflabs.org/timescale/?Ma=443.8

[165] http://tools.wmflabs.org/timescale/?Ma=433.4

[166] Named for the Cefn-Rhuddan Farm in the Llandovery area.

[167] http://tools.wmflabs.org/timescale/?Ma=440.8

[168] http://tools.wmflabs.org/timescale/?Ma=438.5

[169] http://tools.wmflabs.org/timescale/?Ma=427.4

[170] http://tools.wmflabs.org/timescale/?Ma=430.5

[171] http://tools.wmflabs.org/timescale/?Ma=423

[172] http://tools.wmflabs.org/timescale/?Ma=425.6

[173] http://tools.wmflabs.org/timescale/?Ma=419.2

[174] http://stratigraafia.info/materjalid/eesti_strat/Silurian_2015.pdf

[175] , p. 4

[176] https://web.archive.org/web/20120309135815/http://palaeos.com/paleozoic/silurian/silurian.htm

[177] http://www.ucmp.berkeley.edu/silurian/silurian.html

[178] http://www.paleoportal.org/index.php?globalnav=time_space&sectionnav=period&period_id=14

[179] https://web.archive.org/web/20060824004451/http://tapestry.usgs.gov/ages/silurdevon.html

[180] http://www.stratigraphy.org/

[181] http://www.geo-lieven.com/erdzeitalter/silur/silur.htm

[182] http://www.stratigraphy.org/geowhen/stages/Silurian.html

[183] Image:Sauerstoffgehalt-1000mj.svg

[184] File:OxygenLevel-1000ma.svg

[185] Image:Phanerozoic Carbon Dioxide.png

[186] Image:All palaeotemps.png

[187] Rudwick M.S.J. 1985 *The great Devonian controversy: the shaping of scientific knowledge among gentlemanly specialists.* Chicago: University of Chicago Press.

[188] Note: • Sedgwick and Murchison coined the term "Devonian system" in: Adam Sedgwick and Roderick Impey Murchison (1840) "On the physical structure of Devonshire, and on the subdivisions and geological relations of its older stratified deposits, etc.," *Transactions of the Geological Society of London*, 2nd series, **5** (part II) : 633-687 (Part I) and 688-705 (Part II). From p. 701: https://books.google.com/books?id=QknWzPRnVRQC&pg=PA701#v=onepage&q&f=false "We propose therefore, for the future, to designate these groups collectively by the name *Devonian system*, ..." • Sedgwick and Murchison acknowledged William Lonsdale's role in proposing, on the basis of fossil evidence, the existence of a Devonian stratum between those of the Silurian and Carboniferous periods. From (Sedgwick and Murchison, 1840), p. 690: "Again, Mr. Lonsdale, after an extensive examination of the fossils of South Devon, had pronounced them, more than a year since, to form *a group intermediate between those*

of the Carboniferous and Silurian systems, ..." • William Lonsdale stated that in December 1837 he had suggested the existence of a stratum between the Silurian and Carboniferous ones. See: William Lonsdale (1840) "Notes on the age of limestones from south Devonshire," https://books.google.com/books?id=QknWzPRnVRQC&pg=PA721#v=onepage&q&f=false *Transactions of the Geological Society of London*, 2nd series, **5** (part II) : 721-738 ; see especially pp. 724 and 727. From p. 724: " ... Mr. Austen's communication [was] read December 1837, ... It was immediately after the reading of that paper ... that I formed the opinion relative to the limestones of Devonshire being of the age of the old red sandstone; and which I afterwards suggested first to Mr. Murchison and then to Prof. Sedgwick, ..."

[189] Age of Fishes Museum http://www.ageoffishes.org.au

[190] Barclay, W.J. 1989. *Geology of the South Wales Coalfield Pt II, the country around Abergavenny*, 3rd edn. Memoir of the British Geological Survey Sheet 232 (Eng & Wales) pp18-19

[191] http://tools.wmflabs.org/timescale/?Ma=419.2

[192] http://tools.wmflabs.org/timescale/?Ma=393.3

[193] http://tools.wmflabs.org/timescale/?Ma=410.8

[194] http://tools.wmflabs.org/timescale/?Ma=407.6

[195] http://tools.wmflabs.org/timescale/?Ma=387.7

[196] http://tools.wmflabs.org/timescale/?Ma=382.7

[197] http://tools.wmflabs.org/timescale/?Ma=372.2

[198] http://tools.wmflabs.org/timescale/?Ma=358.9

[199]

[200] Devonian Paleogeography http://jan.ucc.nau.edu/~rcb7/devpaleo.html

[201] Benton, M. J. (2005) *Vertebrate Palaeontology* https//books.google.co.nz John Wiley, 3rd edition, page 14.

[202] Palaeos Paleozoic: Devonian: The Devonian Period - 2 http://www.palaeos.com/Paleozoic/Devonian/Devonian.2.htm#Life

[203] C.Michael Hogan. 2010. *Fern*. Encyclopedia of Earth. eds. Saikat Basu and C.Cleveland. National Council for Science and the Environment http://www.eoearth.org/article/Fern?topic=49480. Washington DC.

[204] McGhee, George R. (2013). When the Invasion of Land Failed. New York: Columbia University Press.

[205] Citation needed

[206] http://www.palaeos.com/Paleozoic/Devonian/Devonian.2.htm#Life

[207] http://www.ageoffishes.org.au/

[208] http://www.devoniantimes.org/index.html

[209] http://www.ucmp.berkeley.edu/devonian/devlife.html

[210] http://www.stratigraphy.org/

[211] http://www.geo-lieven.com/erdzeitalter/devon/devon.htm

[212] Image:Sauerstoffgehalt-1000mj.svg

[213] File:OxygenLevel-1000ma.svg

[214] Image:Phanerozoic Carbon Dioxide.png

[215] Image:All palaeotemps.png

[216] Rev. W. D. Conybeare and William Phillips, *Outlines of the Geology of England and Wales* ..., Part I, (London, England: William Phillips, 1822). On page 323, Conybeare titles the chapter "Book III. Medial or Carboniferous Order." https://books.google.com/books?id=ZcMQAAAAIAAJ&pg=PA323#v=onepage&q&f=false

[217] Cossey, P.J. et al. (2004) *British Lower Carboniferous Stratigraphy*, Geological Conservation Review Series, no 29, JNCC, Peterborough (p3)

[218] Menning *et al.* (2006)

[219] Stanley (1999), p 426

[220] Robinson, JM. 1990 Lignin, land plants, and fungi: Biological evolution affecting Phanerozoic oxygen balance. Geology 18; 607–610, on p608.

[221] A HISTORY OF PALAEOZOIC FORESTS - Part 2 The Carboniferous coal swamp forests http://www.uni-muenster.de/GeoPalaeontologie/Palaeo/Palbot/ewald1.htm

[222] C.Michael Hogan. 2010. *Fern*. Encyclopedia of Earth. National council for Science and the Environment http://www.eoearth.org/article/Fern . Washington, DC

[223] Stanley (1999), p 411-12.

[224] M. Alan Kazlev (1998) The Carboniferous Period of the Paleozoic Era: 299 to 359 million years ago http://www.palaeos.com/Paleozoic/Carboniferous/Carboniferous.htm , Palaeos.org, Retrieved on 2008-06-23

[225] Blackwell, Meredith, Vilgalys, Rytas, James, Timothy Y., and Taylor, John W. 2008. Fungi. Eumycota: mushrooms, sac fungi, yeast, molds, rusts, smuts, etc.. Version 21 February 2008. in The Tree of Life Web Project,

[226] Ward, P. et al. (2006): Confirmation of Romer's Gap is a low oxygen interval constraining the timing of initial arthropod and vertebrate terrestrialization. *Proceedings of the National Academy of Sciences* no 103 (45): pp 16818-16822.

[227] http://jeb.biologists.org/cgi/content/abstract/201/8/1043

[228] //doi.org/10.1016/j.palaeo.2006.03.058

[229] https://web.archive.org/web/20060423084018/http://www.stratigraphy.org/gssp.htm

[230] http://www.stratigraphy.org/gssp.htm

[231] https://web.archive.org/web/20130106085716/http://www.stratigraphy.org/bak/geowhen/index.html

[232] http://www.stratigraphy.org/bak/geowhen/index.html

[233] http://www.geo-lieven.com/erdzeitalter/karbon/karbon.htm

[234] http://www.foraminifera.eu/querydb.php?&period=Carboniferous&aktion=suche

[235] Image:Sauerstoffgehalt-1000mj.svg

[236] File:OxygenLevel-1000ma.svg

[237] Image:Phanerozoic Carbon Dioxide.png

[238] Image:All palaeotemps.png

[239] https://www.sciencedaily.com/articles/p/permian-triassic_extinction_event.htm

[240] Benton, M.J. et al., Murchison's first sighting of the Permian, at Vyazniki in 1841 http://palaeo.gly.bris.ac.uk/Benton/reprints/2010Murchison.pdf, Proceedings of the Geologists' Association, accessed 2012-02-21

[241] Murchison, Roderick Impey (1841) "First sketch of some of the principal results of a second geological survey of Russia," https://books.google.com/books?id=1U8wAAAAIAAJ&pg=PA417#v=onepage&q&f=false *Philosophical Magazine and Journal of Science*, series 3, **19** : 417-422. From p. 419: "The carboniferous system is surmounted, to the east of the Volga, by a vast series of marls, schists, limestones, sandstones and conglomerates, to which I propose to give the name of "Permian System," ... ."

[242] "Kazanian" http://www.stratigraphy.org/bak/geowhen/stages/Kazanian.html GeoWhen Database, International Commission on Stratigraphy (ICS)

[243] Palaeos: Life Through Deep Time > The Permian Period http://palaeos.com/paleozoic/permian/permian.htm Accessed 1 April 2013.

[244] Xu, R. & Wang, X.-Q. (1982): Di zhi shi qi Zhongguo ge zhu yao Diqu zhi wu jing guan (Reconstructions of Landscapes in Principal Regions of China). Ke xue chu ban she, Beijing. 55 pages, 25 plates.

[245] Zimmerman EC (1948) Insects of Hawaii, Vol. II. Univ. Hawaii Press

[246] Grzimek HC Bernhard (1975) *Grzimek's Animal Life Encyclopedia* Vol 22 Insects. Van Nostrand Reinhold Co. NY.

[247] Riek EF Kukalova-Peck J (1984) "A new interpretation of dragonfly wing venation based on early Upper Carboniferous fossils from Argentina (Insecta: Odonatoida and basic character states in Pterygote wings.)" *Can. J. Zool.* 62; 1150-1160.

[248] Wakeling JM Ellington CP (1997) Dragonfly flight III lift and power requirements" *Journal of Experimental Biology* 200; 583-600, on p589

[249] Matsuda R (1970) Morphology and evolution of the insect thorax. Mem. Ent. Soc. Can. 76; 1-431.

[250] Riek EF Kukalova-Peck J (1984) A new interpretation of dragonfly wing venation based on early Upper Carboniferous fossils from Argentina (Insecta: Odonatoida and basic character states in Pterygote wings.) Can. J. Zool. 62; 1150-1160

[251] Huttenlocker, A. K., and E. Rega. 2012. The Paleobiology and Bone Microstructure of Pelycosaurian-grade Synapsids. Pp. 90–119 in A. Chinsamy (ed.) Forerunners of Mammals: Radiation, Histology, Biology. Indiana University Press.

341

[252] https://web.archive.org/web/20040219015324/http://www.stratigraphy.org/gssp.htm

[253] http://www.stratigraphy.org/gssp.htm

[254] http://www.ucmp.berkeley.edu/permian/permstrat.html

[255] http://www.ucmp.berkeley.edu/permian/glassmts.html

[256] http://www.stratigraphy.org/

[257] http://www.geo-lieven.com/erdzeitalter/perm/perm.htm

[258] http://www.paleodirect.com/pgset2/amph013.htm

[259] http://geology.gsapubs.org/content/40/3/287.full

[260] http://adsabs.harvard.edu/abs/2012Geo....40..287S

[261] //doi.org/10.1130/focus032012.1

[262] Jones, Daniel (2003) [1917], Peter Roach, James Hartmann and Jane Setter, eds., *English Pronouncing Dictionary*, Cambridge: Cambridge University Press, ISBN 3-12-539683-2

[263] "Mesozoic" http://www.dictionary.com/browse/Mesozoic. *Dictionary.com Unabridged*. Random House.

[264] http://tools.wmflabs.org/timescale/?Ma=251.902–66

[265] http://tools.wmflabs.org/timescale/?Ma=251.902–201.3

[266] http://tools.wmflabs.org/timescale/?Ma=201.3–145

[267] http://tools.wmflabs.org/timescale/?Ma=145–66

[268] See Hughes, T.; " The case for creation of the North Pacific Ocean during the Mesozoic Era" in *Palaeogeography, Palaeoclimatology, Palaeoecology*; Volume 18, Issue 1, August 1975, Pages 1-43

[269] Stanley, Steven M. *Earth System History*. New York: W.H. Freeman and Company, 1999.

[270] http://www.britannica.com/EBchecked/topic/142729/Cretaceous-Period/69972/Paleoclimate

[271] Robert A. Berner, John M. VandenBrooks and Peter D. Ward, 2007, *Oxygen and Evolution* http://www.sciencemag.org/content/316/5824/557.summary. *Science 27 April 2007*, Vol. 316 no. 5824 pp. 557-558 . A graph showing the reconstruction from this paper can be found here http://www.oocities.org/marie.mitchell@rogers.com/climate_files/Phanerozoic_Oxygen.gif, from the webpage Paleoclimate - The History of Climate Change http://www.oocities.org/marie.mitchell@rogers.com/PaleoClimate.html.

[272] Berner R. A. 2006 *GEOCARBSULF: a combined model for Phanerozoic atmospheric O2 and CO2*. Geochim. Cosmochim. Acta 70, 5653–5664. See the dotted line in Fig. 1 of *Atmospheric oxygen level and the evolution of insect body size* http://rspb.royalsocietypublishing.org/content/early/2010/03/04/rspb.2010.0001.full by Jon F. Harrison, Alexander Kaiser and John M. VandenBrooks

[273] Berner, Robert A., 2009, *Phanerozoic atmospheric oxygen: New results using the GEOCARBSULF model* http://www.ajsonline.org/content/309/7/603.abstract. Am. J. Sci. 309 no. 7, 603-606. A graph showing the reconstructed levels in this paper can be found on p. 31 https://books.google.com/books?id=GdRnFn7I38kC&lpg=PP1&pg=PA31#v=onepage&q&f=false of the book *Living Dinosaurs* by Gareth Dyke and Gary Kaiser.

[274] Berner R. A., Canfield D. E. 1989 *A new model for atmospheric oxygen over phanerozoic time*. Am. J. Sci. 289, 333–361. See the solid line in Fig. 1 of *Atmospheric oxygen level and the evolution of insect body size* http://rspb.royalsocietypublishing.org/content/early/2010/03/04/rspb.2010.0001.full by Jon F. Harrison, Alexander Kaiser and John M. VandenBrooks

[275] Berner, R, et al., 2003, *Phanerozoic atmospheric oxygen*, Annu. Rev. Earth Planet. Sci., V, 31, p. 105-134. See the graph near the bottom of the webpage Phanerozoic Eon http://essayweb.net/geology/timeline/phanerozoic.shtml

[276] Glasspool, I.J., Scott, A.C., 2010, *Phanerozoic concentrations of atmospheric oxygen reconstructed from sedimentary charcoal*, Nature Geosciences, 3, 627-630

[277] Bergman N. M., Lenton T. M., Watson A. J. 2004 COPSE: a new model of biogeochemical cycling over Phanaerozoic time. Am. J. Sci. 304, 397–437. See the dashed line in Fig. 1 of *Atmospheric oxygen level and the evolution of insect body size* http://rspb.royalsocietypublishing.org/content/early/2010/03/04/rspb.2010.0001.full by Jon F. Harrison, Alexander Kaiser and John M. VandenBrooks

[278] Stan Baducci. *Mesozoic Plants*. https://archive.is/20130123140455/http://www.fossilnews.com/2000/mezplants/mezplants.html.

[279] C.Michael Hogan. 2010. *Fern*. Encyclopedia of Earth. National council for Science and the Environment http://www.eoearth.org/article/Fern . Washington, DC

[280] Image:Sauerstoffgehalt-1000mj.svg

[281] File:OxygenLevel-1000ma.svg

[282] Image:Phanerozoic Carbon Dioxide.png

[283] Image:All palaeotemps.png

[284] Friedrich von Alberti, *Beitrag zu einer Monographie des bunten Sandsteins, Muschelkalks und Keupers, und die Verbindung dieser Gebilde zu einer Formation* [Contribution to a monograph on the colored sandstone, shell limestone and mudstone, and the joining of these structures into one formation] (Stuttgart and Tübingen, (Germany): J. G. Cotta, 1834). Alberti coined the term "Trias" on page 324 https://books.google.com/books?id=1ClSAAAAcAAJ&pg=PA324#v= onepage&q&f=false :

"... bunter Sandstein, Muschelkalk und Keuper das Resultat einer Periode, ihre Versteinerungen, um mich der Worte E. de Beaumont's zu bedeinen, die Thermometer einer geologischen Epoche seyen, ... also die bis jezt beobachtete Trennung dieser Gebilde in 3 Formationen nicht angemessen, und es mehr dem Begriffe Formation entsprechend sey, sie zu einer Formation, welche ich vorläufig *Trias* nennen will, zu verbinden."

(... colored sandstone, shell limestone, and mudstone are the result of a period ; their fossils are, to avail myself of the words of E. de Beaumont, the thermometer of a geologic epoch ; ... thus the separation of these structures into 3 formations, which has been maintained until now, isn't appropriate, and it is more consistent with the concept of "formation" to join them into one formation, which for now I will name "trias".)

[285]

[286] Stanley, 452-3.

[287] Dott, R.H. and Batten, R.L. (1971) Evolution of the Earth, 4th ed. McGraw Hill, NY.

[288] Darlington PJ, (1965) Biogeography of the southern end of the world. Harvard University Press, Cambridge Mass., on p 168.

[289] Nomade et al.,2007 Palaeogeography, Palaeoclimatology, Palaeoecology 244, 326-344.

[290] Marzoli et al., 1999, Science 284. Extensive 200-million-year-old continental flood basalts of the Central Atlantic Magmatic Province, pp. 618-620.

[291] Hodych & Dunning, 1992.

[292] http://gsa.confex.com/gsa/2002AM/finalprogram/abstract_42936.htm

[293] http://www.stratigraphy.org/bak/gssp.htm

[294] http://inyo.coffeecup.com/site/ammonitecanyon/ammonitecanyon.html

[295] http://inyo.coffeecup.com/site/ammonoids/ammonoids.html

[296] http://inyo.coffeecup.com/site/uw/unionwash.html

[297] http://inyo.coffeecup.com/site/triassic/triassicfossils.html

[298] http://inyo.coffeecup.com/site/thaynes/nevadaammonoids.html

[299] https://web.archive.org/web/20151210044546/http://palaeos.com/mesozoic/triassic/triassic. htm

[300] https://web.archive.org/web/20160222224358/http://rainbow.ldgo.columbia.edu/courses/ v1001/9.html

[301] http://gallery.in-tch.com/~earthhistory/triassic%20page%201.html

[302] http://palaeo.gly.bris.ac.uk/Palaeofiles/Triassic/triextict.htm

[303] http://www.geo-lieven.com/erdzeitalter/trias/trias.htm

[304] Image:Sauerstoffgehalt-1000mj.svg

[305] File:OxygenLevel-1000ma.svg

[306] Image:Phanerozoic Carbon Dioxide.png

[307] Image:All palaeotemps.png

[308] A 140 Ma age for the Jurassic-Cretaceous instead of the usually accepted 145 Ma was proposed in 2014 based on a stratigraphic study of Vaca Muerta Formation in Neuquén Basin, Argentina.<ref>

[309] Alexander von Humboldt, *Kosmos*, volume 4 (Stuttgart: Cotta, 1858), p. 632 http://babel. hathitrust.org/cgi/pt?id=ucm.5322149281;view=1up;seq=638: "Ich hatte mich auf einer geognostischen Reise, die ich 1795 durch das südliche Franken, die westliche Schweiz und Ober-Italien machte, davon überzeugt, daß der Jura-Kalkstein, welchen Werner zu seinem Muschelkalk rechnete, eine eigne Formation bildete. In meiner Schrift über die unterirdischen Gasarten, welche mein Bruder Wilhelm von Humboldt 1799 während meines Aufenthalts in Südamerika herausgab, wird der Formation, die ich vorläufig mit dem Namen Jura-Kalkstein bezeichnete, zuerst (S. 39) gedacht." ('On a geological tour that I made in 1795 through southern France, western Switzerland and upper Italy, I convinced myself that the Jura limestone, which Werner included in his shell limestone, constituted a separate formation. In my paper about subterranean types of gases, which my brother Wilhelm von Humboldt published in 1799 during my stay in South America, the formation, which I provisionally designated with the name "Jura limestone", is first conceived (p. 39).'

[310] Alexander von Humboldt, *Ueber die unterirdischen Gasarten und die Mittel, ihren Nachteil zu vermindern, ein Beitrag zur Physik der praktischen Bergbaukunde* ['On the types of subterranean gases and means of minimizing their harm, a contribution to the physics of practical mining'] (Braunschweig: Vieweg, 1799), p. 39 https://books.google.com/books?id= xiQ7AAAAcAAJ&pg=PA39#v=onepage&q&f=false: "[...] die ausgebreitete Formation, welche zwischen dem alten Gips und neueren Sandstein liegt, und welchen ich vorläufig mit dem Nahmen Jura-Kalkstein bezeichne." ('... the widespread formation which lies between the old gypsum and the more recent sandstone and which I provisionally designate with the name "Jura limestone".')

[311] Hölder, H. 1964. *Jura – Handbuch der stratigraphischen Geologie*, IV. Enke-Verlag, Stuttgart.

[312] Arkell, W.J. 1956. *Jurassic Geology of the World*. Oliver & Boyd, Edinburgh und London.

[313] Pieńkowski, G.; Schudack, M.E.; Bosák, P.; Enay, R.; Feldman-Olszewska, A.; Golonka, J.; Gutowski, J.; Herngreen, G.F.W.; Jordan, P.; Krobicki, M.; Lathuiliere, B.; Leinfelder, R.R.; Michalík, J.; Mönnig, E.; Noe-Nygaard, N.; Pálfy, J.; Pint, A.; Rasser, M.W.; Reisdorf, A.G.; Schmid, D.U.; Schweigert, G.; Surlyk, F.; Wetzel, A. & Theo E. Wong, T.E. 2008. "Jurassic". In: McCann, T. (ed.): *The Geology of Central Europe*. Volume 2: *Mesozoic and Cenozoic*, Geological Society, London, pp. 823–922.

[314] Rollier, L. 1903. *Das Schweizerische Juragebirge* (Sonderabdruck aus dem Geographischen Lexikon der Schweiz). Verlag von Gebr. Attinger, Neuenburg.

[315] Late Jurassic http://www.scotese.com/late1.htm

[316] Monroe and Wicander, 607.

[317]

[318] Motani, R. (2000), Rulers of the Jurassic Seas, Scientific American vol.283, no. 6

[319] Why the giant azhdarchid Arambourgiania philadelphiae needs a fanclub http://markwitton-com.blogspot.pt/2016/06/why-giant-azhdarchid-arambourgiania.html

[320] Haines, 2000.

[321] Behrensmeyer *et al.*, 1992, 349.

[322] Behrensmeyer *et al.*, 1992, 352

[323] Behrensmeyer *et al.*, 1992, 353

[324] https://web.archive.org/web/20060105125654/http://www.palaeos.com/Mesozoic/Jurassic/Jurassic.htm

[325] http://adsabs.harvard.edu/abs/1998PPP...144....3S

[326] //doi.org/10.1016/s0031-0182%2898%2900109-6

[327] https://web.archive.org/web/20090325233234/http://www.wooster.edu/geology/Taylor%26Wilson2003.pdf

[328] http://adsabs.harvard.edu/abs/2003ESRv...62....1T

[329] //doi.org/10.1016/s0012-8252%2802%2900131-9

[330] http://www3.wooster.edu/geology/Taylor%26Wilson2003.pdf

[331] http://www.geo-lieven.com/erdzeitalter/jura/jura.htm

[332] https://web.archive.org/web/20081208011604/http://harbury.villagebuzz.co.uk/viewtopic.php?f=16&t=297

[333] http://www.foraminifera.eu/querydb.php?period=Jurassic&aktion=suche

[334] Image:Sauerstoffgehalt-1000mj.svg

[335] File:OxygenLevel-1000ma.svg

[336] Image:Phanerozoic Carbon Dioxide.png

[337] Image:All palaeotemps.png

[338] From page 373: "La troisième, qui correspond à ce qu'on a déja appelé formation de la craie, sera désigné par le nom de terrain crétacé." (The third, which corresponds to what was already called the "chalk formation", will be designated by the name "chalky terrain".)

[339] The term "Cretaceous" first appeared in English in: Henry Thomas De La Beche, *A Geological Manual* (Philadelphia, Pennsylvania: Carey & Lea, 1832). From page 35: https://books.google. com/books?id=nb0QAAAAIAAJ&pg=PA35#v=onepage&q&f=false "Group 4. (*Cretaceous*) contains the rocks which in England and the North of France are characterized by chalk in the upper part, and sands and sandstones in the lower."

[340] See Stanley (1999), pp. 481–482

[341] Dougal Dixon et al., *Atlas of Life on Earth*, (New York: Barnes & Noble Books, 2001), p. 215.

[342] Stanley, Steven M. *Earth System History.* New York: W.H. Freeman and Company, 1999. p. 280

[343] Stanley, pp. 279–81

[344] Nordt, Lee, Stacy Atchley, and Steve Dworkin. "Terrestrial Evidence for Two Greenhouse Events in the Latest Cretaceous." Baylor University, 6 Oct. 2003. Web. 10 Nov. 2012.

[345] Stanley, pp. 480–2

[346] Stanley, pp. 481–2

[347] "Warmer than a Hot Tub: Atlantic Ocean Temperatures Much Higher in the Past" http://www. physorg.com/news10978.html PhysOrg.com. Retrieved 12/3/06.

[348] Skinner, Brian J., and Stephen C. Porter. *The Dynamic Earth: An Introduction to Physical Geology.* 3rd ed. New York: John Wiley & Sons, Inc., 1995. p. 557

[349] C.Michael Hogan. 2010. *Fern.* Encyclopedia of Earth. National council for Science and the Environment http://www.eoearth.org/article/Fern . Washington, DC

[350] Kielan-Jaworowska, Zofia, Richard L. Cifelli, and Zhe-Xi Luo (2005). Mammals from the Age of Dinosaurs: Origins, Evolution, and Structure, p. 299

[351] https://www.sciencedirect.com/science/article/pii/S0146638007002628

[352] //doi.org/10.1016/j.orggeochem.2007.11.010

[353] https://web.archive.org/web/20060824202146/http://www.ucm.es/info/estratig/vol31/10ovek. pdf

[354] http://www.ucm.es/info/estratig/vol31/10ovek.pdf

[355] http://adsabs.harvard.edu/abs/2003ESRv...62....1T

[356] //doi.org/10.1016/S0012-8252%2802%2900131-9

[357] http://www.ucmp.berkeley.edu/mesozoic/cretaceous/cretaceous.html

[358] https://web.archive.org/web/20121201000325/http://www3.wooster.edu/geology/bioerosion/ bioerosion.html

[359] http://www.foraminifera.eu/querydb.php?period=Cretaceous&aktion=suche

[360] "Cenozoic" http://www.dictionary.com/browse/Cenozoic. *Dictionary.com Unabridged.* Random House.

[361] "Cenozoic" https://www.merriam-webster.com/dictionary/Cenozoic. *Merriam-Webster Dictionary.*

[362] "Cainozoic" http://www.dictionary.com/browse/Cainozoic. *Dictionary.com Unabridged.* Random House.

[363] Oxford English Dictionary Second Edition, 1989

[364] From pp. 153–154: "As many systems or combinations of organic forms as are clearly traceable in the stratified crust of the globe, so many corresponding terms (as Palæozoic, Mesozoic, Kainozoic, &c.) may be made, ... "

[365] International Stratigraphic Chart http://www.stratigraphy.org/ICSchart/ChronostratChart2014-02.pdf

[366] http://tools.wmflabs.org/timescale/?Ma=55–45

[367] http://tools.wmflabs.org/timescale/?Ma=35

[368] http://tools.wmflabs.org/timescale/?Ma=55.5

[369] http://tools.wmflabs.org/timescale/?Ma=49

[370] http://tools.wmflabs.org/timescale/?Ma=41

[371] http://tools.wmflabs.org/timescale/?Ma=2.8

[372] http://museum.wa.gov.au/explore/dinosaur-discovery/age-mammals

[373] Image:Sauerstoffgehalt-1000mj.svg

[374] File:OxygenLevel-1000ma.svg

[375] Image:Phanerozoic Carbon Dioxide.png

[376] Image:All palaeotemps.png

[377] Formerly, the period covered by the Paleogene was called the first part of the Tertiary, whose usage is no longer official. "Whatever happened to the Tertiary and Quaternary?" http://www.stratigraphy.org/bak/geowhen/TQ.html

[378] Robert W. Meredith, Jan E. Janecka, John Gatesy, Oliver A. Ryder, Colleen A. Fisher, Emma C. Teeling, Alisha Goodbla, Eduardo Eizirik, Taiz L. L. Simão, Tanja Stadler, Daniel L. Rabosky, Rodney L. Honeycutt, John J. Flynn, Colleen M. Ingram, Cynthia Steiner, Tiffani L. Williams, Terence J. Robinson, Angela Burk-Herrick, Michael Westerman, Nadia A. Ayoub, Mark S. Springer, William J. Murphy. 2011. Impacts of the Cretaceous Terrestrial Revolution and KPg extinction on mammal diversification. Science 334:521-524.

[379] http://www.foraminifera.eu/querydb.php?period=Paleogene&aktion=suche

[380] http://tools.wmflabs.org/timescale/?Ma=66–56

[381] Hooker, J.J., "Tertiary to Present: Paleocene", pp. 459-465, Vol. 5. of Selley, Richard C., L. Robin McCocks, and Ian R. Plimer, Encyclopedia of Geology, Oxford: Elsevier Limited, 2005.

[382] Stephen Jay Gould, ed., The Book of Life (New York: W.W. Norton & Company, 1993), p. 182.

[383] Kazlev, M. Alan (2002) " The Paleocene" http://palaeos.com/cenozoic/paleocene/paleocene.htm. Palaeos Cenozoic. Retrieved April 3, 2013.

[384] https//www.researchgate.net

[385] Linnéa Smeds, Hans Ellegren, The Dynamics of Incomplete Lineage Sorting across the Ancient Adaptive Radiation of Neoavian Birds http://journals.plos.org/plosbiology/article?id=10.1371/journal.pbio.1002224

[386] http://www.stratigraphy.org/gssp.htm

[387] http://paleocene-mammals.de/

[388] https://web.archive.org/web/20050306074829/http://www.bbc.co.uk/beasts/changing/paleocene/index.shtml

[389] http://www.itano.net/fossils/marylan2/marylan2.htm

[390] https://web.archive.org/web/20041204140749/http://www.palaeos.com/Cenozoic/Paleocene/Paleocene.htm

[391] http://www.nceas.ucsb.edu/~alroy/Paleocene.html

[392] http://scotese.com/

[393] http://www.foraminifera.eu/querydb.php?age=Paleocene&aktion=suche

[394] http://petrifiedwoodmuseum.org/PaleoceneIntroduction.htm

[395] http://paleobiology.si.edu/geotime/main/htmlversion/paleocene1.html

[396] http://tools.wmflabs.org/timescale/?Ma=56–33.9

[397] The extinction of the Hantkeninidae, a planktonic family of foraminifera became generally accepted as marking the Eocene-Oligocene boundary; in 1998 Massignano in Umbria, central Italy, was designated the Global Boundary Stratotype Section and Point (GSSP).

[398] See: • Letter from William Whewell to Charles Lyell dated 31 January 1831 in: • From p. 55: "The period next antecedent we shall call Eocene, from ήως, aurora, and χαινος, recens, because the extremely small proportion of living species contained in these strata, indicates what may be considered the first commencement, or dawn, of the existing state of the animate creation."

[399] http://anthro.palomar.edu/earlyprimates/early_2.htm

[400] Myhre, G., D. Shindell, F.-M. Bréon, W. Collins, J. Fuglestvedt, J. Huang, D. Koch, J.-F. Lamarque, D. Lee, B. Mendoza, T. Nakajima, A. Robock, G. Stephens, T. Takemura and H. Zhang (2013) "Anthropogenic and Natural Radiative Forcing" http://www.climatechange2013.org/images/report/WG1AR5_Chapter08_FINAL.pdf. In: Climate Change 2013: The Physical Science Basis. Contribution of Working Group I to the Fifth Assessment Report of the Intergovernmental Panel on Climate Change. Stocker, T.F., D. Qin, G.-K. Plattner, M. Tignor, S.K.

Allen, J. Boschung, A. Nauels, Y. Xia, V. Bex and P.M. Midgley (eds.). Cambridge University Press, Cambridge, United Kingdom and New York, NY, USA.

[401] http://www.scotese.com/

[402] https://web.archive.org/web/20041010134818/http://www.palaeos.com/Cenozoic/Eocene/Eocene.htm

[403] https://www.pbs.org/wgbh/evolution/change/deeptime/eocene.html

[404] http://www.dmap.co.uk/fossils/

[405] https://web.archive.org/web/20051116075532/http://www.sas.upenn.edu/earth/arctic/

[406] https://web.archive.org/web/20090531203432/http://members.cox.net/pyrophyllite/Basilosaurus1.html

[407] http://www.oceansofkansas.com/Harlan1834b.html

[408] http://jan.ucc.nau.edu/~rcb7/terpaleo.html

[409] http://www.scotese.com/newpage9.htm

[410] http://www.foraminifera.eu/querydb.php?age=Eocene&aktion=suche

[411] Haines, Tim; *Walking with Beasts: A Prehistoric Safari*, (New York: Dorling Kindersley Publishing, Inc., 1999)

[412] http://tools.wmflabs.org/timescale/?Ma=28.1

[413] http://tools.wmflabs.org/timescale/?Ma=23.03

[414] http://tools.wmflabs.org/timescale/?Ma=33.9

[415] A.Zanazzi (et al.) 2007 'Large Temperature Drop across the Eocene Oligocene in central North America' Nature, Vol. 445, 8 February 2007

[416] C.R.Riesselman (et al.) 2007 'High Resolution stable isotope and carbonate variability during the early Oligocene climate transition: Walvis Ridge (ODPSite 1263) USGS OF-2007-1047

[417] Lorraine E. Lisiecki Nov 2004; *A Pliocene–Pleistocene stack of 57 globally distributed benthic $\delta^{18}O$ records* Brown University, PALEOCEANOGRAPHY, VOL. 20

[418] Kenneth G. Miller Jan–Feb 2006; *Eocene–Oligocene global climate and sea-level changes St. Stephens Quarry, Alabama* GSA Bulletin, Rutgers University, NJ http://bulletin.geoscienceworld.org/cgi/reprint/120/1-2/34?eaf

[419] http://palaeos.com/cenozoic/oligocene/oligocene.html

[420] http://www.ucmp.berkeley.edu/tertiary/oli.html

[421] http://www.copyrightexpired.com/earlyimage/index.html

[422] http://www.paleodirect.com/pl-003.htm

[423] http://www.paleodirect.com/fl-001.htm

[424] http://scotese.com/oligocen.htm

[425] http://www.foraminifera.eu/querydb.php?age=Oligocene&aktion=suche

[426] Image:Sauerstoffgehalt-1000mj.svg

[427] File:OxygenLevel-1000ma.svg

[428] Image:Phanerozoic Carbon Dioxide.png

[429] Image:All palaeotemps.png

[430] The term "Neogene" was coined in 1853 by the Austrian palaeontologist Moritz Hörnes (1815–1868): From p. 806: *"Das häufige Vorkommen der Wiener Mollusken ... im trennenden Gegensatze zu den eocänen zusammenzufassen."* (The frequent occurrence of Viennese mollusks in typical Miocene as well as in typical Pliocene deposits motivated me – in order to avoid the perpetual monotony [of providing] details about the deposits – to subsume both deposits provisionally under the name "Neogene" (νεος new and γιγνομαι to arise) in distinguishing contrast to the Eocene.)

[431] Tucker, M. E. (2001) *Sedimentary Petrology* (3rd ed.) Blackwell Science, Osney Nead, Oxford, UK,

[432] Lourens, L., Hilgen, F., Shackleton, N.J., Laskar, J., Wilson, D., (2004) "The Neogene Period". In: Gradstein, F., Ogg, J., Smith, A.G. (Eds.), *Geologic Time Scale* Cambridge University Press, Cambridge.

[433] Clague, John et al. (2006) "Open Letter by INQUA Executive Committee" http://www.inqua.tcd.ie/documents/QP%2016-1.pdf *Quaternary Perspective, the INQUA Newsletter* International Union for Quaternary Research 16(1)

[434] https://web.archive.org/web/20130423191403/http://geosun.sjsu.edu/~jhendricks/AtlasTemp/neogene.html

[435] "Miocene" http://www.dictionary.com/browse/Miocene. *Dictionary.com Unabridged*. Random House.

[436] "Miocene" https://www.merriam-webster.com/dictionary/Miocene. *Merriam-Webster Dictionary*.

[437] http://tools.wmflabs.org/timescale/?ma=23.03–5.333

[438] See: • Letter from William Whewell to Charles Lyell dated 31 January 1831 in: • From p. 54: "The next antecedent tertiary epoch we shall name Miocene, from μειων, minor, and χαινος, recens, a minority only of fossil shells imbedded in the formations of this period, being of recent species."

[439] http://tools.wmflabs.org/timescale/?Ma=7.5–5.6

[440] https://web.archive.org/web/20060423084018/http://www.stratigraphy.org/gssp.htm

[441] https://www.pbs.org/wgbh/evolution/change/deeptime/miocene.html

[442] http://www.ucmp.berkeley.edu/tertiary/mio.html

[443] http://www.foraminifera.eu/querydb.php?age=Miocene&aktion=suche

[444] http://humanorigins.si.edu/evidence/human-evolution-timeline-interactive

[445] "Pliocene" https://www.merriam-webster.com/dictionary/Pliocene. *Merriam-Webster Dictionary*.

[446] "Pliocene" http://www.dictionary.com/browse/Pliocene. *Dictionary.com Unabridged*. Random House.

[447] "Pleiocene" http://www.dictionary.com/browse/Pleiocene. *Dictionary.com Unabridged*. Random House.

[448] See the 2014 version of the ICS geologic time scale http://www.stratigraphy.org/ICSchart/ChronostratChart2014-02.jpg

[449] See: • Letter from William Whewell to Charles Lyell dated 31 January 1831 in: • From p. 53: "We derive the term Pliocene from πλειων, major, and χαινος, recens, as the major part of the fossil testacea of this epoch are referrible to recent species*."

[450] Van Andel (1994), p. 226.

[451] Comins & Kaufmann (2005), p. 359.

[452] https://web.archive.org/web/20060423084018/http://www.stratigraphy.org/gssp.htm

[453] http://www.stratigraphy.org/gssp.htm

[454] https://web.archive.org/web/20050623075602/http://www.giss.nasa.gov/research/features/pliocene/

[455] https://web.archive.org/web/20041207033036/http://www.palaeos.com/Cenozoic/Pliocene/Pliocene.htm

[456] https://www.pbs.org/wgbh/evolution/change/deeptime/pliocene.html

[457] http://gsa.confex.com/gsa/2007AM/finalprogram/abstract_129643.htm

[458] https://web.archive.org/web/20060221061156/http://www.sciencenews.org/articles/20020202/fob5.asp

[459] http://www.ucmp.berkeley.edu/tertiary/pli.html

[460] http://www.foraminifera.eu/querydb.php?age=Pliocene&aktion=suche

[461] http://humanorigins.si.edu/evidence/human-evolution-timeline-interactive

[462] Image:Sauerstoffgehalt-1000mj.svg

[463] File:OxygenLevel-1000ma.svg

[464] Image:Phanerozoic Carbon Dioxide.png

[465] Image:All palaeotemps.png

[466] Earthquake Glossary - Late Quaternary https://earthquake.usgs.gov/learn/glossary/?term=Late%20Quaternary U.S. Geological Survey

[467] See: • Available at: Museo Galileo (Florence (Firenze), Italy) https://bibdig.museogalileo.it/Teca/Viewer;jsessionid=3160EE486323866602B9590459D7111B?an=323812_6 From p. 158 (clviii): *"Per quanto ho potuto sinora osservavare, la serie di questi strati, che compongono la corteccia visibile della terra, mi pare distinta in quattro ordini generali, e successivi, senza considerarvi il mare."* (As far as I have been able to observe, the series of these layers that compose the visible crust of the earth seems to me distinct in four general orders, and successive, not considering the sea.) • English translation:

[468] From p. 193: https://www.biodiversitylibrary.org/item/29350#page/199/mode/1up *"Ce que je désirerais ... dont il faut également les distinguer."* (What I would desire to prove above all is

that the series of tertiary deposits continued – and even began in the more recent basins – for a long time, perhaps after that of the Seine had been completely filled, and that these later formations – *Quaternary* (1), so to say – should not retain the name of alluvial deposits any more than the true and ancient tertiary deposits, from which they must also be distinguished.) However, on the very same page, Desnoyers abandoned the use of the term "quaternary" because the distinction between quaternary and tertiary deposits wasn't clear. From p. 193: *'La crainte de voir mal comprise ... que ceux du bassin de la Seine.'* (The fear of seeing my opinion in this regard be misunderstood or exaggerated, has made me abandon the word "quaternary", which at first I had wanted to apply to all deposits more recent than those of the Seine basin.)

[469] See the 2009 version of the ICS geologic time scale http://www.quaternary.stratigraphy.org.uk/correlation/GSAchron09.jpg

[470] http://www.quaternary.stratigraphy.org.uk/

[471] http://www.quaternary.stratigraphy.org.uk/correlation/chart.html

[472] http://www.quaternary.stratigraphy.org.uk/correlation/POSTERSTRAT_BOREAS_v2005c.pdf

[473] http://www.quaternary.stratigraphy.org.uk/correlation/POSTERSTRAT_v2007b_small.jpg

[474] https://web.archive.org/web/20120626140850/http://tierra.rediris.es/aequa/doc/tabla_aequav2_2009.pdf

[475] http://ice.tsu.ru/index.php?view=category&catid=27&option=com_joomgallery&Itemid=40

[476] *In* Gradstein, F. M., Ogg, James G., and Smith, A. Gilbert (eds.), *A Geologic Time Scale 2004* Cambridge University Press, Cambridge,

[477] The Middle Pleistocene and Upper Pleistocene are actually subseries/subepochs rather than stages/ages but, in 2009, the IUGS decided to replace each of them with a stage/age.<ref name="ICSSQS">

[478] From p. 621: *'Toutefois, en même temps ... et de substituer à la dénomination de Nouveau Pliocène celle plus abrégée de* Pleistocène, *tirée du grec pleiston, plus, et kainos, récent.'* (However, at the same time that it became necessary to subdivide the two periods mentioned above, I found that the terms intended to designate these subdivisions were of an inconvenient length, and I have proposed to use in the future the word "Pliocene" for "old Pliocene", and to substitute for the name "new Pliocene" this shorter "Pleistocene", drawn from the Greek *pleiston* (most) and *kainos* (recent).)

[479] For the top of the series, see:

[480] Riccardi, Alberto C. (30 June 2009) "IUGS ratified ICS Recommendation on redefinition of Pleistocene and formal definition of base of Quaternary" http://www.stratigraphy.org/upload/IUGS%20Ratification_Q%20&%20Pleistocene.pdf International Union of Geological Sciences

[481] Gradstein, Felix M.; Ogg, James G. and Smith, A. Gilbert (eds.) (2005) *A Geologic Time Scale 2004* Cambridge University Press, Cambridge, UK, p. 28,

[482] National Geographic Channel, *Six Degrees Could Change The World,* Mark Lynas interview. Retrieved February 14, 2008.

[483] Roy, M., P.U. Clark, R.W. Barendregt, J.R., Glasmann, and R.J. Enkin, 2004, *Glacial stratigraphy and paleomagnetism of late Cenozoic deposits of the north-central United States* http://geo.oregonstate.edu/files/geo/Royetal-GSAB-2004.pdf, PDF version, 1.2 MB. Geological Society of America Bulletin.116(1-2): pp. 30-41;

[484] (contains a summary of how and why the Nebraskan, Aftonian, Kansan, and Yarmouthian stages were abandoned by modern stratigraphers).

[485] http://siberiantimes.com/science/casestudy/news/worms-frozen-in-permafrost-for-up-to-42000-years-come-back-to-life

[486] Stringer, C.B. (1992) "Evolution of early modern humans" *In*: Jones, Steve; Martin, R. and Pilbeam, David R. (eds.) (1992) *The Cambridge encyclopedia of human evolution* Cambridge University Press, Cambridge, , pp. 241–251.

[487] http://www.stratigraphy.org/gssp.htm

[488] http://digitalcollections.smu.edu/cdm/ref/collection/sit/id/23

[489] http://www.foraminifera.eu/querydb.php?age=Pleistocene&aktion=suche

[490] https/www.academia.edu

[491] http://humanorigins.si.edu/evidence/human-evolution-timeline-interactive

[492] //en.wikipedia.org/w/index.php?title=Template:Holocene&action=edit

[493] Formal subdivision of the Holocene Series/Epoch https://www.qpg.geog.cam.ac.uk/news/formalsubdivisionoftheholoceneseriesgeogr18.pdf

[494] The name "Holocene" was proposed in 1850 by the French palaeontologist and entomologist Paul Gervais (1816–1879): From p. 413: https://babel.hathitrust.org/cgi/pt?id=mdp.39015012332287;view=1up;seq=453 *"On pourrait aussi appeler* Holocènes, *ceux de l'époque historique, ou dont le dépôt n'est pas antérieur à la présence de l'homme ; ... "* (One could also call "Holocene" those [deposits] of the historic era, or the deposit of which is not prior to the presence of man ; ...)

[495] http://www.bbc.co.uk/nature/history_of_the_earth/Holocene

[496] http://ghkclass.com/ghkC.html?Holocene

Article Sources and Contributors

The sources listed for each article provide more detailed licensing information including the copyright status, the copyright owner, and the license conditions.

Geological history of Earth *Source:* https://en.wikipedia.org/w/index.php?oldid=843100522 *License:* Creative Commons Attribution-Share Alike 3.0 *Contributors:* 72, Adam9007, Alfie Gandon, Altenmann, Amortias, Amp71, AndrewHowse, Andycjp, Aunt Entropy, Avenue, Awickert, Awsome2212, Azuris, BD2412, BSATwinTowers, Bazonka, Bear-rings, Bejnar, Bender235, Bettymnz4, Bilalnaik, Bongwarrior, Burwellian, Cadiomals, Canthusus, Catalaalatac, Chimesmonster, Chris the speller, Citation bot 1, ClueBot NG, Colonies Chris, CommonsDelinker, Craic Den, CuriousMind01, D.M.N., DSachan, Dawn Bard, Dcirovic, Deadbeef, Deor, Der Golem, DerBorg, Doug Weller, Drbogdan, Dwaipayanc, Eastlaw, ErgoSum88, Excirial, Froth, GeoWriter, Geologyguy, Gilliam, GoingBatty, Greenpac4mp, Hamsterlopithecus, Hardyplants, Harryboyles, Headbomb, Hmains, Hughesyjj, Isambard Kingdom, IznoRepeat, J 1982, J-stan, JLincoln, Jack Greenmaven, Jon Kolbert, JonRichfield, Jpvandijk, KP Botany, Khazar, Kintaro, Kozuch, LeadSongDog, Leovizza, Leszek Jańczuk, LittleOldMe, Lugia2453, MCdance101, Materialscientist, McZusatz, Metiscus, Michael Devore, Micheletb, Mikenorton, Mikeo, Muhends, Mwtoews, NatureA16, NawlinWiki, Neito Nossal, Non-dropframe, NotWith, Novangelis, Onel5969, Onetruepurple, Originalbigj, Oshwah, Parsa, Paul H., Peter M. Brown, Pinethicket, Preciousjfm, R'n'B, RJHall, Rcsprinter123, ReeceStewart, Rjwilmsi, RockMagnetist, Rocket000, Sairjohn∼enwiki, Serendipodous, SpaceChimp1992, Spamalot360, Srich32977, Stinger20, Suicidalhamster, Sushant gupta, The Thing That Should Not Be, The Transhumanist, TheCascadian, Tide rolls, Tom.Reding, UBeR, Viriditas, Vsmith, WereSpielChequers, Widr, Yamara, Yngvadottir, 154 anonymous edits 1

Precambrian *Source:* https://en.wikipedia.org/w/index.php?oldid=851953521 *License:* Creative Commons Attribution-Share Alike 3.0 *Contributors:* Acroterion, AlexiusHoratius, Alfie Gandon, AlphaBetaGamma01, Anaxial, Anna Frodesiak, Arado, Attilios, Ayman3322, Bdyawhduid, BigDwiki, Bollyjeff, Booduhbaby, Boombangbang, Bri, CAPTAIN RAJU, Cephal-odd, Chander, ClueBot NG, DaEliteGamer123, Donner60, Dpleibovitz, Drbogdan, Dud1010, Ehrenkater, El C, Emilyneblock, Fadesga, Fama Clamosa, Fireypepper, FunkMonk, Gap9551, GeoWriter, Geopersona, Gilliam, Giraffedata, GoodBoyInc, Gothicsea, GünniX, Hacapella, Hairy Dude, Hello71, Heronbarracuda, Horseless Headman, Humphrey Jungle, IAreC4, IronGargoyle, Isambard Kingdom, JKshaw, Jabberwoch, Jbeans, Jjkdog88, Joestape89, JorisvS, Josve05a, Julietdeltalima, Jwsanniota, Keithicus420, Kintetsubuffalo, KylieTastic, Libitz, Look2See1, LuK3, Lugia2453, Mandruss, Manofwar53, Member, Metaloaf, Mike1000000000, Mikenorton, Mindmatrix, MusikAnimal, Napzilla, Neogene252, Omnipaedista, Oshwah, PigeonOfTheNight, Plantsurfer, Proxima Centauri, RA0808, RandomLittleHelper, Red Planet X (Hercolubus), Redgro, Renamed user sdfkjlskdfreu8r98, Renniksdb, Rhododendrites, Rjdotorg, Rmosler2100, RockMagnetist, Rockisy, Russell Mason7, Semmendinger, Serols, Sky4903, Slade, Space Infinite, Suede Cat, Sweepy, Tajotep, Teleohapsis, Tentinator, TerryAlex, Theslomoguy128, Tom.Reding, Trackteur, Trevayne08, V620 Cephei, Volcanoguy, Vsmith, WNYY98, Wikipelli, Wilson44691, Zedshort, 187 anonymous edits 19

Hadean *Source:* https://en.wikipedia.org/w/index.php?oldid=852367263 *License:* Creative Commons Attribution-Share Alike 3.0 *Contributors:* Abductive, Abyssal, Ai.protopapas, Akjar13, Alansohn, Andrewb686, Andycjp, Anyeverybody, Ashmoo, Axl, Bender235, Bento00, BlaiseFEgan, Bmorrisett, BobIsToBobasBackIsToBlack, Bobamnertiopsis, Bobrayner, Brendanconway, Bri, CRGreathouse, Cadiomals, Captainbeefart, CheeseSqueezy, Cherkash, Chermundy, Chris Weimer, Citation bot 1, ClueBot NG, Colonel64, CommonsDelinker, Courcelles, Cph3992, Crs666, Crum375, DVdm, Daiyusha, Darylgolden, Deflective, Dger, Dicksuk, DinosaursLoveExistence, Donner60, Drbogdan, Dunkleosteus77, Emijrp, EnglishWoodsman, Epicgenius, Esefsfag, Eteq, Excirial, F Notebook, Fama Clamosa, Fig wright, Fred Bauder, Gdestroyer66, GeoWriter, Geopersona, Gfoley4, Gilliam, Glane23, GravisZro, GrumpyFromBrighton, Hairy Dude, Hamish59, Headbomb, Heron, Hibernian, Hiyabulldog, Hojunchang, Hydrogeology, Iadmc, Iblardi, IceKarma, Igodard, Jacobisdumb1234, JaconaFrere, Jan1nad, Jay Heins, Jbaranao, Jbeans, Jbening, Jdfekete, Jim1138, Jimp, Jni, Jntg4Games, Jquarry, Klilidiplomus, Koavf, Kwamikagami, Lankenau, Lankiveil, Leptictidium, Lithopsian, Look2See1, Lycurgus, Magioladitis, MarioProtIV, MatthewHoobin, Mike Rosoft, Mikenorton, Monty845, Mully9, NawlinWiki, Nightscream, Nihiltres, Omnipaedista, Optakeover, Orangebrickfalls, Orangemarlin, PCock, Palaeocene, Panyd, Peter Karlsen, Pevernagie, Peyre, Phatency, Philip Trueman, Pppery, PranksterTurtle, Prokaryotes, Ravenswing, Red Planet X (Hercolubus), ReddyHakky1998, Retired user 0001, Rjwilmsi, Roentgenium111, Rursus, SamX, SamiAEH, Scientific29, Scwlong, Sdevoid, SmallMossie, Snow Blizzard, Steve Quinn, The PIPE, TheNeutroniumAlchemist, Tide rolls, Tim1357, Tobias1984, Tom.Reding, TreyAlfonso, Triangulum, Tyler021398, V620 Cephei, VexorAbVikipædia, Volcanoguy, Vsmith, Waso99, Wikipedian 2, X201, You Don't Do That!, Zedshort, Zsinj, \wowzeryest\, 174 anonymous edits ...27

Archean *Source:* https://en.wikipedia.org/w/index.php?oldid=850386928 *License:* Creative Commons Attribution-Share Alike 3.0 *Contributors:* 17C ext2015, A. Parrot, A2soup, Abyssal, AdjustShift, Ahecht, Alansohn, Alpertron, Alumnum, Andrewjlockley, ApokryItaroes, Apokryltaros, Ataleh, Aurora2698, Axl, Bender235, Buddyboy76, Cadiomals, Casualsentencestructureandsciencerectifier, Chermundy, Citation bot 1, ClueBot NG, Crum375, DARTH SIDIOUS 2, DHN-bot∼enwiki, David1010, DavidBrooks-AWB, Deflective, Dino, DinosaursLoveExistence, Dodo von den Bergen, Drbogdan, Dudley Miles, Dunkleosteus77, Edward, Ekren, EmersonLowry, Eteethan, Eumolpo, Excirial, Eyreland, F Notebook, Fabartus, Fabrictramp, Falstromkat, Francvs, Gap9551, Geekdiva, GeoWriter, Geologyguy, Geopersona, Gob Lofa, Gorthian, Graeme Bartlett, Hater70769, Headbomb, HedAurabesh, Hike395, Hojunchang, I dream of horses, Igel 14, Jayron32, Jdfekete, John Cline, KaplanAL, Katalaveno, Khalid Mahmood, Kwamikagami, Looie496, Look2See1, Lovkal, Lugia2453, Luk, MaeseLeon, Magioladitis, MarioProtIV, Martinor, Materialscientist, MatthewDBA, Mdsam2∼enwiki, Mikenorton, Modest Genius, Msdkmk, Narndt, Niceguyedc, Offnfopt, OrenBochman, Oshwah, Ozielke, Peter Karlsen, Phe, Pinethicket, PlyrStar93, Profjack, Ptbotgourou, RA0808, Racingstripes, Rasnaboy, Razorflame, Red Planet X (Hercolubus), Red58bill, Rich Farmbrough, Rjwilmsi, Robin S, Robot433, RockMagnetist, Romit3, Rursus, SMcCandlish, Seaphoto, Serols, Sl8roof, Smith609, Snow Blizzard, Solarra, Srich32977, Ssolbergj, Sushant gupta, The Thing That Should Not Be, Thuvan Dihn, Tom Prangnell, Tom.Reding, Tompop888, Trappist the monk, Trevayne08, Triangulum, Twinsday, Unrepetant, Vanished User jdksfajlasd, VexorAbVikipædia, VoABot II, Volcanoguy, Vsmith, Waso99, Welisongered, Widefox, WikiMasterGhibif, WolfmanSF, Yamara, Yortzec, Zedshort, 154 anonymous edits 33

Proterozoic *Source:* https://en.wikipedia.org/w/index.php?oldid=852358014 *License:* Creative Commons Attribution-Share Alike 3.0 *Contributors:* 0xF8E8, AMK152, Ahoerstemeier, Alansohn, AlefZet, Alumnum, Andrewjlockley, Anthony Appleyard, Archaeodontosaurus, Ardric47, Are you ready for IPv6?, Arjayay, Arkuat, Ascánder, Attilios, Awickert, Barbara Shack, Bejnar, Bobamnertiopsis, Boneyard90, Bryan Derksen, CLCStudent, Cadiomals, Casito, Cenarium, Chermundy, Chochopk, Chris the speller, ClueBot NG, Cyan, DHN-bot∼enwiki, Da Joe, DadaNeem, Danilot, DatGuy, Davidweman, Dcirovic, Deflective, DegreeofGlory, DinoBenn, DocWatson42, Donarreiskoffer, Dougofborg, Drbogdan, Dreadstar, Drmies, Emperorbma, Enlil Ninlil, Eras-mus, EricGrandillo, Erimus, Fama Clamosa, Fang 23, Fanghong∼enwiki, FoCuSandLeArN, Gene.arboit, GeoWriter, Geologyguy, Gerhardvalentin, Gilliam, Giraffedata, Glenn, Gorank4, Gregmb81, Haitao Huang, Happysailor, Hike395, Horsebrutality, I dream of horses, IvanLanin, J. 'mach' wust, Japanese Searobin, Jay Gregg, Jc-S0CO, Jimp, Joe2719, John, Joy, Jsonitsac, Jyril, Kalogeropoulos, Kevmin, Kku, Kwamikagami, Kww, Kyle1278, LilHelpa, Look2See1, Lugia2453, MTSbot∼enwiki, Mahewa, MarioProtIV, Marto217127, Mikenorton, Mindmatrix, Miss Madeline, Moe Epsilon, MountainB, Moverton, MrJohnnyMorales, Muneeb Khaar, My Chemistry romantic, Myasuda, NatureA16, NeoJustin, NewEnglandYankee, Nicolas Perrault III, Nihiltres, NimbusWeb, Nwbeeson, OatmealFTLOTUDO, Oddbodz, P.Geol, Pengo, Petrb, Peyre, Phe, Philip Trueman, Plantsurfer, Poolkris, Qxz, RHB, Red Planet X (Hercolubus), Rettetast, Rjwilmsi, Robin S, Rocket000, Ryan Shakiba, SamX, SkyLined, Slawojarek, Smith609, Srich32977, Steinbach, Suede Cat, Sushant gupta, TaBOT-zerem, Tarodar, Tarquin, TerraFrost, The Rambling Man, The Thing That Should Not Be, Toyokuni3, Uanfala, Unyoyega, Videogameplayer99, Volcanoguy, Vsmith, Vuong Ngan Ha, Waso99, Wilson44691, Wingedsubmariner, Xihuzhaiyue, Xionbox, Zedshort, Zeno Gantner, Šedý, [illegible], 144 anonymous edits 38

Phanerozoic *Source:* https://en.wikipedia.org/w/index.php?oldid=853550087 *License:* Creative Commons Attribution-Share Alike 3.0 *Contributors:* AMK152, Abyssal, Alansohn, Alfie Gandon, Alnagov, Alpha xp5, Alphathon, Anaxial, Apokryltaros, Arskoul, Artemis-Arethusa, AshLin, BD2412, BarretB, Bcasterline, Bejnar, Bender235, BenoniBot∼enwiki, Boydstra, Br77rino, Cadiomals, Cameta, Chermundy, Chipmunkdavis, Clinamental, ClueBot NG, ColRad85, DHN-bot∼enwiki, Danilot, DatGuy, David Latapie, David1010, Dawnseeker2000, Dcirovic, Deflective, Dekimasu, Derek R Bullamore, DocWatson42, Donarreiskoffer, Dragons flight, Drbogdan, Dunkleosteus77, Ehrenkater, Elassint, Enirac Sum, Eog1916, Epipelagic, Eras-mus, Filemon, Floria L, Fryed-peach, Gazibara, Gee Eight, GeoWriter, Glenn, Gorank4, Graham87, Grahbudd, Guy Macon, Hairy Dude, Headbomb, Henrychianski, Hike395, Ibrahim Husain Meraj, Igor Topilsky, Isambard Kingdom, IvanLanin, Ivgnyl, J. 'mach' wust, Jc-S0CO, Jdabney, Jhbdel, John of Reading, JohnWheater, JorisvS, Jrundin, Jumbuck, Jyril, Kazubon∼enwiki, Kku, Kwamikagami, L Kensington, Lee Vilenski, Leonard G., Look2See1, MTSbot∼enwiki, MacHaddock, Mani1, Metanoid, Mikenorton, Modest Genius, Mutamarrid, Myasuda, NTBot∼enwiki, Oangola, Obsidian Soul, Olend444, OrenBochman, Oxymoron83, P Aculeius, Palica, Patentpedant, Petr Matas, Piledhigheranddeeper, Plantdrew, Purbo T, R'n'B, Rcsprinter123, Rich Farmbrough, Ringbang, RockMagnetist, Saturn comes back around, SchreiberBike, Serols, Sesu Prime, Siim, Smith609, Smjg, Sofia Koutsouveli, Sushant gupta, TStein, TaBOT-zerem, Tevildo, The Palaeontologist, TheNeutroniumAlchemist, Tobias1984, Tom.Reding, Topbanana, Ugog Nizdast, VexorAbVikipædia, Vsmith, Vuong Ngan Ha, Winterst, Xerxes314, Yamara, Yangula, Zedshort, ,'דקי [illegible], 141 anonymous edits 45

Paleozoic *Source:* https://en.wikipedia.org/w/index.php?oldid=850347533 *License:* Creative Commons Attribution-Share Alike 3.0 *Contributors:* 4lextintor, 72, Abyssal, Alex Cohn, Alfie Gandon, Amorymeltzer, Anaxial, Arjayay, Ayoitzquan, BD2412, Beta7, Bruce Lee's Relative "Chin", Cadiomals, CambridgeBayWeather, CheckYouser, ChrisGualtieri, Classicwiki, Clinamental, ClueBot NG, Codiv, Crystallizedcarbon, D.Lazard, Danger, DavidLeighEllis, Dcirovic, Deflective, Discospinster, Donner60, Dudley Miles, Dunkleosteus77, Ehrenkater, El C, Emeraldcityserendipity, Eurodyne, Excirial, Exercisephys, Fama Clamosa, Finnusertop, Flyer22 Reborn, Fredricwilliams, Freyzian, Gap9551, Gareth Griffith-Jones, Gazelle55, GeoWriter, Geopersona, GermanJoe, Gilgamesh∼enwiki, Gilliam, Glossologist, H.dryad, Hike395, InedibleHulk, IronGargoyle, Isambard Kingdom, J. 'mach' wust, Jamesx12345, Jarble, Jdfekete, Jeffrd10, Jenks24, Jiten D, Justinclined11, Katieh5584, Kelisi, Kittycat60, Kwamikagami, Littlelittlethgs, Look2See1, Lugia2453, Magioladitis, Materialscientist, Melcous, Mikehadlow, Mikenorton, Mngreg, Mononomic, MrJohnnyMorales, MusikAnimal, Naturebrain1, Neropwnsu, Newyorkadam, NitroBum, Obsidian Soul, Ogress, OmegaBuddy13, Oshwah, Pinethicket, Plantdrew, Plantsurfer, Poyekhali, Quietbritishjim,

Qzd, RainFall, Rcsprinter123, Robert K S, Rockut, Run27i, STH235SilverLover, Seaphoto, Secretkeeper12, Serols, Shellwood, Sjö, Skamecrazy123, SkyWarrior, Someone9901, Spumuq, Stesmo, TCN7JM, TYelliot, Telfordbuck, Thanatos666, Thefjordhusky201, Thejoepeach, Thirdright, Tom.Reding, Username Needed, VexorAbVikipædia, Vsmith, Wikipelli, XenoVon, Yintan, Zedshort, 206 anonymous edits 62

Cambrian *Source:* https://en.wikipedia.org/w/index.php?oldid=853224498 *License:* Creative Commons Attribution-Share Alike 3.0 *Contributors:* 0xF8E8, 72, A. B., A3X2, AManWithNoPlan, Abyssal, AnonMoos, AtticTapestry, BBKoVI, Bamyers99, Bender235, BetterSkatez, Boneyard90, Bong-warrior, CAPTAIN RAJU, CASSIOPEIA, CLCStudent, Cadiomals, Capitalismojo, CheckYouser, Chese hole, Chiswick Chap, Chrimas1, ChrisGualtieri, ClueBot NG, CommonsDelinker, CuriousEric, Daicaregos, Daniel Patterson98, Dcirovic, DemocraticLuntz, Dinoguy2, Diverman, Drbogdan, Epicgenius, Excedron9238, Excirial, Fama Clamosa, Gap9551, GeoWriter, Gilgamesh~enwiki, Gorobay, Gorthian, Graphium, Grisgm, Headbomb, Hike395, I am the one the one, I dream of horses, Isambard Kingdom, Jason Quinn, Jfmantis, Jim1138, Johaniseenkneus, JorisvS, Josve05a, Julietdeltalima, Justanonymous, Justinejeff28, Kazvorpal, Kennethcgass, Kintaro, Kurzon, Kwamikagami, Lappspira, Lavalizard101, Leo Fyllnet, Libitz, Loganfalco, Look2See1, Lugia2453, Marasama, Marek69, MarioProtIV, Megatherium, Mikenorton, Moist towelett, MusikAnimal, Mys 721tx, Niceguyedc, Nick Number, O.Koslowski, Obsidian Soul, Odawg101, Ohconfucius, Omnipaedista, Orphan Wiki, Oshwah, OwenBlacker, PBS-AWB, Paul venter, Pdcook, Peace Makes Plenty, Philip Trueman, Pinethicket, Pkbwcgs, Plantsurfer, PlyrStar93, Pokeman918, QQuantum, Rambam rashi, Rcsprinter123, Renamed user sdfkjlskdfreu8r98, Retallack, Rich Farmbrough, Rjwilmsi, RockMagnetist, Roentgenium111, Serols, Shirudo, Smalljim, Smith609, StoryKai, Tamarack velocity, The Letter J, Tobias1984, Trappist the monk, TropicalCyclones243, Twsx, VEO15, VexorAbVikipædia, Vieque, Vsmith, Wavelength, Widr, Wikimedes, Wikishovel, Wilson44691, Yung hacka, Zedshort, Zollo9999, Zzyzx11, Érico, 151 anonymous edits 71

Ordovician *Source:* https://en.wikipedia.org/w/index.php?oldid=853574690 *License:* Creative Commons Attribution-Share Alike 3.0 *Contributors:* A. Parrot, Abyssal, Anaxial, Asdfjkl1235, Auseplot, BD2412, Bender235, Bkell, Cadiomals, Chermundy, Chris Capoccia, Chris the speller, Citation bot 1, ClueBot NG, CommonsDelinker, Couchtater44, DARTH SIDIOUS 2, DaL33T, DanielCD, Dawnseeker2000, Deflective, Dimitrii, Discospinster, Download, Dwergenpaartje, Edderso, Ehrenkater, Eluchil404, Eniagrom, Excedron9238, Excirial, Fama Clamosa, Flyer22 Reborn, Gemini1980, GeoWriter, Geoffrey.landis, Geopersona, Ginsuloft, Giorgiogp2, Gulumeemee, GünniX, Harelx, Hike395, HrishikeshShidhaye, JJuano, Jack Greenmaven, Jc3s5h, Jjknack1, Joe2719, JoseBonner, Justinejeff28, K6ka, Kent G. Budge, Kurzon, Kwamikagami, Lappspira, Littlelittlethgs, Livitup, Look2See1, MTWEmperor, Marek69, MarioProtIV, Martarius, Materialscientist, Meganesia, MelbourneStar, Melilac, Metaldev, Mgiganteus1, Michael C Price, Mike Rosoft, Mikenorton, Mild Bill Hiccup, Mys 721tx, Necessary Evil, Neko-chan, Nerds alike, NewEnglandYankee, NotWith, Obsidian Soul, Olev Vinn, Onel5969, Orcoteuthis, Pachyteuthis, Pepper, PiPieGuy, Pigsonthewing, Plantsurfer, PlyrStar93, Pratyya Ghosh, R'n'B, RA0808, Rcsprinter123, Red Planet X (Hercolubus), Ribakau, Rjwilmsi, Robocon1, RockMagnetist, RomanSpa, SamX, Samf4u, ScotsmaninUtah, Septrillion, Sirnature, Skizzik, Smith609, Snowolf, Stemonitis, StoryKai, TAnthony, TRowlette, The Anonymouse, The Utahraptor, Theodomas, Tide rolls, Tobias1984, Topbanana, Toyokuni3, Vader119, VexorAbVikipædia, Vkil, Vsmith, Wetman, Wikipelli, WillSchenk, Wilson44691, Wjfox2005, WoodPig, Zedshort, Zzyzx11, 161 anonymous edits .. 86

Silurian *Source:* https://en.wikipedia.org/w/index.php?oldid=851624285 *License:* Creative Commons Attribution-Share Alike 3.0 *Contributors:* 9Questions, A. Parrot, Abyssal, Alan G. Archer, Anbu121, Animalparty, Archon 2488, Argovian, BDD, Badobadop, Bender235, Bgwhite, BurtAlert, Bwtranch, CLCStudent, Cadiomals, Calvinteah96, Canthusus, Chris Capoccia, ClueBot NG, CommonsDelinker, DVdm, Dan56, Dan653, DatGuy, Dd72499, Discospinster, DivineAlpha, DocWatson42, Dohn joe, Download, Drbogdan, Dudley Miles, DuncanHill, EditWikipedia, Elijah.Hall, Epbr123, EuroCarGT, Excedron9238, Fama Clamosa, Florian Blaschke, Flyer22 Reborn, GeoWriter, Geopersona, Gilliam, Glane23, Global Cerebral Ischemia, Gluons12, Haploidavey, Hike395, Horseless Headman, I dream of horses, JaconaFrere, Jauhienij, Justinclined11, Justinejeff28, Kaleb Schmidt, KaplanAL, Kendrick7, Khalid Mahmood, Kirbanzo, Kittenono, Klbrain, Kurzon, Kwamikagami, La Noia de Barcelona, Lavalizard101, Littlelittlethgs, Look2See1, Lysippos, Magog the Ogre, Manofwar53, Marek69, MarioProtIV, Melilac, Michael C Price, Mikenorton, Mr. Billion, MusikAnimal, Myasuda, Nepenthes, O0fishguy0o, Olev Vinn, Oshwah, Peter coxhead, Petter Bøckman, Philip Trueman, Plantsurfer, Propagandaprince, Quinton Feldberg, Qwertyuiop1994, RA0808, Red Planet X (Hercolubus), RichardWeiss, Rjwilmsi, RockMagnetist, Rrburke, Sanyi4, Serols, ShadowRangerRIT, SkyLined, Slightsmile, Smalljim, Sophie means wisdom, Sven Manguard, TAnthony, Tarheel95, TerryAlex, Thincat, Tide rolls, Tisquesusa, Tobias1984, Tolly4bolly, Tom Morris, Tombanks1234, UY Scuti, Ubiquity, Ueutyi, Vermont, Vinnyzz, Vsmith, Wayne Hardman, Wbm1058, Widefox, Widr, Wiki13, Wilson44691, Yintan, Zedshort, Ζετα ζ, 182 anonymous edits 107

Devonian *Source:* https://en.wikipedia.org/w/index.php?oldid=852374189 *License:* Creative Commons Attribution-Share Alike 3.0 *Contributors:* 37ophiuchi, 90 Auto, Abyssal, Alan G. Archer, Alaney2k, Altaïr, Anaxial, Anticline, Apidium23, Apokryltaros, Applehatersinthehouse, Audaciter, BD2412, BethNaught, Bgwhite, Cadiomals, Camillo123, ChrisGualtieri, Clean Copy, ClueBot NG, CommonsDelinker, Csigabi, DVdm, DavidBrooks-AWB, Dawn Bard, Dbachmann, Discospinster, DocWatson42, Drbogdan, Drewmutt, Ducknish, Epipelagic, Ferrdann, FoCuSandLeArN, Fuortu, Gap9551, GeoWriter, Ghuhani, Gilliam, Ginsuloft, Glacialfox, Gogo Dodo, GünniX, Hairisfun, HapHaxion, Hike395, Idenshi, Jaberwocky888, Jakec, Jenks24, Jim1138, Jjkdog88, Jschnur, Justinejeff28, Kenyon, Klilidiplomus, Kurzon, Lavalizard101, Learningthroughlife, Lionhead99, Look2See1, Lugia2453, Luxure, Lythronaxargestes, Macman252, Magog the Ogre, MarioProtIV, Mark Arsten, Materialscientist, Mcfar54, Mchanges!, Meganesia, Metacomet10, Mgiganteus1, Mikenorton, Minimac, Miracle Pen, Mogism, Mys 721tx, NewEnglandYankee, Nick Wilson, Nwbeeson, Obsidian Soul, Oerjan, Ogress, Pablomartinez, Parunach, Petter Bøckman, Peyre, Philip Trueman, Pinethicket, Plantsurfer, Prokaryotes, Qbgeekjtw, Rambam rashi, Rawr2015, RichardWeiss, Rjwilmsi, RockMagnetist, SHFW70, SMB99thx, SamX, Seagullqueen, SeminulaOolite, Serols, Shanzad, Shellwood, Slazenger, Smalljim, Sophie means wisdom, Stevietheman, Stingwicki, TAnthony, Tall-timothy, The Triple M, Thirdright, Thnidu, Tmangibsonsscience, ToBeFree, Tobias1984, Toddst1, Trappist the monk, Trollol3000500, Trollol3000600, Turnspten, TwoTwoHello, Uanfala, Valenciano, Varlaam, Vertium, VexorAbVikipædia, Vlogdud101, Vsmith, WadeSimMiser, Wavelength, Wikipelli, Wilson44691, Yintan, Zedshort, Zzuuzz, Zzyzx11, বাবা করণ, Ὁ οἶστρος, 166 anonymous edits 119

Carboniferous *Source:* https://en.wikipedia.org/w/index.php?oldid=852723549 *License:* Creative Commons Attribution-Share Alike 3.0 *Contributors:* A. Parrot, A2soup, Abyssal, Alphathon, Animalparty, Attilios, AxelBoldt, BD2412, Bahudhara, BallenaBlanca, Bender235, BiT, Bob Burkhardt, Bposert, BrightStarSky, CLCStudent, Cadiomals, Careful With That Axe, Eugene, ChrisGualtieri, ClueBot NG, CommonsDelinker, Corinne, Cwkmail, DatGuy, David.moreno72, Dcirovic, Deli nk, Diademodon, DivermanAU, DuncanHill, Dyanega, Dylankyle04, Euna Asyncritus, Excirial, Fadesga, Fafnir1, Fama Clamosa, Finetooth, Fluffernutter, Flyer22 Reborn, FlyingAce, Gap9551, GeoWriter, Geopersona, GoShow, Gorthian, Hayman30, Headbomb, Hegurka, Hellow ww my nam is joh eiughw4piufdk, Hike395, Igel 14, IronGargoyle, Isambard Kingdom, Jackfork, James086, Jarble, Jbeans, Jdlawlis, Jenks24, Jetstreamer, Jgamleus, Jim1138, JohannSnow, JonRichfield, Jonathan Tweet, Justin15w, Justinejeff28, KN2731, Kempf EK, Kevin12xd, Kibi78704, Kku, Kurzon, Lavalizard101, Lbdearmas, Lightlowemon, Look2See1, MZMcBride, Magioladitis, MarioProtIV, Martarius, Materialscientist, Me, Myself, and I are Here, Meganesia, Mhese, Mikenorton, Myasuda, Mys 721tx, NHSavage, NerdyScienceDude, Netloc, Niceguyedc, Noyster, Nyttend, Onore Baka Sama, Oshwah, Paleocemoski, Plantsurfer, R'n'B, REBECCAE7, Rallette, Redrose64, Reedman72, Rex4680, Rjwilmsi, Robert650, RockMagnetist, Rrburke, Rygel, M.C., Saintrain, SamX, Seaphoto, SemiHypercube, SheriffIsInTown, Simplexity22, SkyWarrior, Smalegander5, Smartypantsssss12, Squinge, Steinbach, SupernovaExplosion, TAnthony, Taylordw, The Proffesor, TheSuave, Tom.Reding, Treisijs, Vagr7, Vsmith, W.stanovsky, Wasell, Wavelength, Werieth, Widr, Wilson44691, Yintan, Yolo69er, Zedshort, Zzyzx11, Σ, אֵבְרָהָם, 174 anonymous edits 135

Permian *Source:* https://en.wikipedia.org/w/index.php?oldid=852718710 *License:* Creative Commons Attribution-Share Alike 3.0 *Contributors:* Abyssal, AddWittyNameHere, Ahaigh9877, Akmedtheterrorist, Altaïr, Annexatious, Anonymous from the 21st century, Ashphalt, BD2412, Bahudhara, Bfedward13, Brickhouse187, Brickhouse777, Brickwahl69, CLCStudent, Cadiomals, Cleric2145, ClueBot NG, Colonies Chris, CommonsDelinker, Corinne, DVdm, Darwin's Bulldog, Dcirovic, Demon 123, DerHexer, Dinoguy2, Donner60, Drewmutt, Dudley Miles, Dunkleosteus77, Edgar181, Ehrenkater, El C, Euna Asyncritus, Excirial, Fama Clamosa, Flyer22 Reborn, Fyrael, Gap9551, GeoWriter, Geoffrey.landis, Gilliam, Guthix20, HapHaxion, Hello71, Heron, Hike395, IAmAPersonMan, IShadowed, Ijon, Imperator Nox, Ira Leviton, IronGargoyle, Jab843, Jaguar, James Carter1, Jennica, Jim1138, Joao Xavier, Joserk, Justinejeff28, Kehkou, Khalid Mahmood, King2360, Klow, Kmg90, Kostja, Kris159, Kurzon, L-Bit, Lavalizard101, Levelbaconman123, Littlelittlethgs, Look2See1, Lotje, Lythronaxargestes, Magioladitis, Majora4, MarioProtIV, Martarius, Martinor, Mean as custard, Mikenorton, Mollykane8, Monado, MusikAnimal, Mys 721tx, Nellington, Non-dropframe, Ogress, Orangemarlin, PCHS-NJROTC, Paleocemoski, Pinethicket, PlyrStar93, Postdlf, QuartierLatin1968, Qzd, Regulov, Rjwilmsi, RockMagnetist, Sagginkcuf, Sedman87, Sergio Kaminski, Serols, Shadowjams, Shellwood, Spencer, StarmanW, Stedil, Swegmasterson, Te Karere, Tentinator, Thelivinghorror, Therizinosaurian, Tobias1984, Toyokuni3, Treisijs, Trusilver, Unclevinny, Velella, VexorAbVikipædia, Vrenator, Vsmith, WOSlinker, Webclient101, Wesley J M, Widr, Wilson44691, WolfmanSF, Wtmitchell, XJonah12K, Zedshort, Zzyzx11, 166 anonymous edits 158

Mesozoic *Source:* https://en.wikipedia.org/w/index.php?oldid=850512102 *License:* Creative Commons Attribution-Share Alike 3.0 *Contributors:* A.amitkumar, Abyssal, AddWittyNameHere, Alfie Gandon, AmericanAir88, Anmol Shrivastav, Apokryltaros, Arjayay, AsceticRose, Ashphalt, Bender235, Boboforchocho, Bogobor, Bradleymcphearson, CASSIOPEIA, Cadiomals, Chaoyangopterus, Chrisyoung1974, ClueBot NG, ColRad85, Dan653, Daniel.Cardenas, DatGoodDude342, DavidLeighEllis, Dawnseeker2000, Dayshade, Dcirovic, Deflective, Deor, Dicklyon, Dl2000, Doge miner se, Donner60, Dunkleosteus77, EAST JACKSON88888, El C, Evangelos Giakoumatos, EvergreenFir, Excirial, Falconfly, Fama Clamosa, Finnusertop, Firemario1130, Freemo1999, Frietjes, Gap9551, GeoDude111, GeoWriter, Gilliam, Headbomb, Hike395, I am One of Many, Ichthyovenator, IronGargoyle, J. 'mach' wust, J. Spencer, Joserk, Just a guy from the KP, Kaapitone, Katieh5584, Kirbanzo, Kwamikagami, KylieTastic, Linguist111, Lionbryce1, Look2See1, Lugia2453, Macman252, Mandruss, Materialscientist, Mikenorton, Mind(a)Maker, Mkmcconn, Moosehadley, MusikAnimal, Nick Moyes, O.Koslowski, Optakeover, Oshwah, Peter M. Brown, Pinethicket, Plaba123, Plantdrew, RA0808, Racerx11, Rain drop 45, Rcsprinter123, STH235SilverLover, SamX, Secretkeeper12, Serols, Stephenb, SwissChez, TAnthony, TheMesquito, There'sNoTime, Thewolrab, Tom.Reding, Topbanana, TropicalCyclones243, VS6507, Velella, Verbal.noun, VexorAbVikipædia, Vsmith, WolfmanSF, 205 anonymous edits 171

Triassic *Source:* https://en.wikipedia.org/w/index.php?oldid=852633141 *License:* Creative Commons Attribution-Share Alike 3.0 *Contributors:* 10metreh, 123ezekielt, A. Parrot, Abyssal, Acroterion, AmanSinghSamra, Animalparty, Anmol Shrivastav, Apokryltaros, AtticTapestry, Attilios, Bangerz23, Beland, Bender235, Bentogoa, Besieged, Bgwhite, BobEnyart, Bordwall, Breadlord69, Brickhouse777, C.Logan, CLCStudent, Caftaric, Clanksided, ClueBot NG, CommonsDelinker, Cwkmail, Danny Sprinkle, Dawnseeker2000, Dcirovic, Deflective, Dillard421, Dominicanpapi82, Donner60, El C, Eric-Wester, Ethanlu121, Eyesnore, Fafnir1, Falconfly, FallingGravity, Flyer22 Reborn, FreeKnowledgeCreator, Frietjes, Frosty, GB fan, Gap9551, Gareth Griffith-Jones, Gatemansgc, GeoWriter, Giants2012, Gilliam, Headbomb, Hike395, Home Lander, I love turtles, Italia2006, J. Spencer, James086, JamesRobert271, Jason Quinn, Jemmasimmons, Jni, Jonesey95, Justinejeff28, K. Badri Vishal, Kaapitone, KaplanAL, King jakob c, Kku, Kmm, Kurzon, Kwamikagami, Lappspira, Lavalizard101, Liz, Look2See1, Lourdes, LuigiPortaro29, MRD2014, Mandruss, MarioProtIV, Martarius, Meganesia, Melody

Concerto, Mikenorton, Mild Bill Hiccup, Muktaka Joshipura, Mys 721tx, Narky Blert, Neptune's Trident, NewEnglandYankee, Niceguyedc, Now in3D, Oczwap, Optakeover, PaleoGeekSquared, Peter M. Brown, Pinethicket, Polaris·Ursa, Qfl247, Rcsprinter123, Rhinestone K, Rjwilmsi, Rmosler2100, Robevans123, RockMagnetist, Rodw, Rsrikanth05, Ruakh, SHFW70, SNUGGUMS, Sadads, Sergio Kaminski, SheriffIsInTown, Simplexity22, Slightsmile, Soap, The Anomebot2, Thirdright, ThomasMikael, Tobias1984, Tolly4bolly, TomS TDotO, Trappist the monk, TuttiFruttiCherryPie, TwoTwoHello, Vsmith, WNYY98, Widr, Wilson44691, Yggdrasilius, Zedshort, Zzyzx11, 179 anonymous edits 182

Jurassic *Source:* https://en.wikipedia.org/w/index.php?oldid=853556718 *License:* Creative Commons Attribution-Share Alike 3.0 *Contributors:* 72, Abyssal, Allthefoxes, AnotherNewAccount, Archaeopsittacus, Bender235, Bermicourt, Big universe, Blaylockjam10, BlueFenixReborn, Brettr21, CLCStudent, CX42, Cadillac000, ChouxMonster, ClueBot NG, CommonsDelinker, CopernicusAD, DVdm, Darorcilmir, Dawei20, Dcirovic, DemocraticLuntz, Dinosaur Cove Dreaming, Discospinster, DivineAlpha, DocWatson42, Donner60, Drewmutt, Drrock, Ehrenkater, El C, Elliot Teong, Excedron9238, Excirial, Falconfly, Florian Blaschke, Flyer22 Reborn, GB fan, GFraggz, Gadget850, Gap9551, GeoWriter, Gilliam, Gob Lofa, Heartwood2003, Hike395, Hughesdavidw, Hydrargyrum, IiKkEe, IronGargoyle, Ixfd64, J 1982, J. Spencer, Jax 0677, Jie Wu, Julietdeltalima, Junior5a, Justinejeff28, Kikipol, Krenair, KylieTastic, Look2See1, Macman252, Maczkopeti, Magere Hein, Malcolm77, Mandruss, Marasama, MarioProtIV, Materialscientist, Matt7899, Mayty1221, Meganesia, Mikenorton, My name is not dave, Mys 721tx, NewEnglandYankee, Newk14, Oleaster, Olev Vinn, Oshwah, Person who formerly started with "216", Petrb, Philip Trueman, Pinethicket, Pleiotrop3, QuartierLatin1968, Rjwilmsi, RyanTQuinn, SHLaowai, Sagginkcuf, Serols, Shellwood, Sietecolores, Simplexity22, Smalljim, Strike Eagle, Super48paul, T-Rex girl, Titus III, Tylercollins 10, VexorAbVikipædia, Vsmith, Wetman, Widr, Wiki13, Wikipediapotato, Wilson44691, Wjfox2005, Wobrey, Yewtharaptor, Yung hacka, ZLEA, Zedshort, Zouke20, 253 anonymous edits 199

Cretaceous *Source:* https://en.wikipedia.org/w/index.php?oldid=851391896 *License:* Creative Commons Attribution-Share Alike 3.0 *Contributors:* 72, Akmedtheterrorist, Alessandro57, Alfie Gandon, Anmol Srivastav, Apokryltaros, Atlanta Jones, Ayil676, B14709, BD2412, Basicdesign, Bender235, Bogmontyvill, C.Fred, CaveStory, Chaoyangopterus, ClueBot NG, Coasterlover1994, CommonsDelinker, Coviekiller5, Daniel.Cardenas, Davidbird764, Dawnseeker2000, Dcirovic, Ddead, Dewritech, Dicklyon, DivermanAU, Donner60, Drewmutt, Drmies, El C, Excirial, Falconfly, FrB.TG, Frankman101, Gap9551, GeoWriter, Geopersona, Gilliam, Gorthian, HammerFilmFan, HandsomeFella, Happysailor, Headbomb, Hegurka, Hike395, Ilber8000, IronGargoyle, J 1982, J. 'mach' wust, J. Spencer, JFG, Jason Quinn, Jfraatz, Josve05a, Julietdeltalima, Justinejeff28, KAP03, KP Botany, Kevmin, Khajidha, Kirbanzo, KylieTastic, Lincoln Josh, Lindenhurst Liberty, Look2See1, LuigiPortaro29, Macman252, Mandruss, MarioProtIV, Mariomassone, Martarius, Marvellous Spider-Man, Materialscientist, Mean as custard, Meganesia, Mikenorton, Mild Bill Hiccup, Mr bossy, MusikAnimal, Mys 721tx, Mz7, Nardog, Ncmvocalist, NewEnglandYankee, Notapheonix, Orphan Wiki, Oshwah, PaleCloudedWhite, Petrb, Pinethicket, Prestonmag, Rich Farmbrough, Rjwilmsi, RockMagnetist, Rsrikanth05, STH235SilverLover, Sabarina Weng, Secretkeeper12, Serols, Shellwood, Shuvuuia, Sietecolores, Smd75jr, Spizaetus, Theinstantmatrix, Thejocraft, Traveling Man, TwoTwoHello, Tymon.r, Ueutyi, VexorAbVikipædia, Vsmith, WelcometoJurassicPark, Werieth, Widr, Wilson44691, Zedshort, 187 anonymous edits 214

Cenozoic *Source:* https://en.wikipedia.org/w/index.php?oldid=852958981 *License:* Creative Commons Attribution-Share Alike 3.0 *Contributors:* A520, Abyssal, Adam9007, Alfie Gandon, Anaxial, Antiqueight, Apidium23, Assmucher3000, Asukite, BeenAroundAWhile, Bigjudetoo, Biografer, Bronze2018, CLCStudent, ClueBot NG, Cnwilliams, CommonsDelinker, Crystallizedcarbon, DVdm, DanielRigal, DavidLeighEllis, Dawnseeker2000, Dcirovic, Discospinster, DocWatson42, Dragon guy, Driftwoodzebulin, Dunkleosteus77, El C, Eleassar, Esquivalience, Fafnir1, Fraggle81, FreeKnowledgeCreator, Freemo1999, Gap9551, GeoWriter, Giant2236, Gorthian, GünniX, Hedja, Hike395, Ibrahim Husain Meraj, Isambard Kingdom, J. 'mach' wust, Jack Greenmaven, Jackcarey245, Jasmin Ros, Jbeans, JimVC3, Jperrylsu, KH-1, Kangaroolou, Katherinev WAM, Kbseah, Knife-in-the-drawer, Kolbasz, Kwamikagami, KylieTastic, Lacypaperclip, Lappspira, Look2See1, Lugia2453, Lythronax93, MONGO, Macman252, Maczkopeti, Mark Arsten, MasterWikiMan1129, Materialscientist, Mattman00000, Mikenorton, Miranche, Mkdw, Mlkj, Moemomooomo, MusikAnimal, Nechemia Iron, Neliboys3, Niceguyedc, Ocdgrammarian, Onel5969, Oshwah, Ost316, Paleocemoski, Pbsouthwood, PericlesofAthens, Peter Delmonte, Philip Trueman, Pjvandehaar, Racerx11, Rain drop 45, Rcsprinter123, Rich Farmbrough, Risk34, Rjwilmsi, Ryan 9548, Serols, Sfan00 IMG, Sro23, TAnthony, Thehoopisonfire, Titanoshark, TomS TDotO, Tompop888, Unbuttered Parsnip, VexorAbVikipædia, Vk.alberta, Vorziblix, Vsmith, WadeSimMiser, WikiSnippets, Wikipelli, Wywin, Xirostar2002, Yamaguchi先生 , ZfJames, Zzuuzz, 206 anonymous edits 233

Paleogene *Source:* https://en.wikipedia.org/w/index.php?oldid=846372213 *License:* Creative Commons Attribution-Share Alike 3.0 *Contributors:* 37ophiuchi, 83d40m, AMK152, Abyssal, Acabashi, AlefZet, Animalparty, Arjayay, Beagel, Bejnar, Bourguinon, Bryan Derksen, Cadiomals, Casito, Chermundy, ClueBot NG, Colincbn, CommonsDelinker, Conversion script, Crocadog, Crystallizedcarbon, DN-boards1, DVdm, Deflective, Denton22, Deor, Derek Ross, Discospinster, Don Kenney, Drmies, Ekren, El C, Eluchil404, Emperorbma, Eras-mus, Excedron9238, Eyu100, FT2, Fama Clamosa, Fig wright, Finnusertop, Firsfron, FlyingLeopard2014, Fossiliferous, Frietjes, GcSwRhIc, GeoWriter, Geologyguy, Geopersona, Glenn, Graham87, Gulumeemee, Hairy Dude, Henriettemarie159, Hike395, Intoronto1125, IvanLanin, Ixfd64, J. 'mach' wust, J.delanoy, Janet1983, Jarble, Jc-S0CO, JorisvS, Joy, Jsonitsac, Justinejeff28, Jyril, Korovioff, Kurzon, Kwamikagami, KylieTastic, Leszek Jańczuk, Lofor, Lollerwaffle, Look2See1, LordEnder64, MTSbot~enwiki, Macronyx~enwiki, Maczkopeti, MadGuy7023, Mar vin kaiser, MarioProtIV, Mark Arsten, Mav, Mejor Los Indios, MelbourneStar, Mensch, Mhese, Michael C Price, Mikenorton, Miss Madeline, Mys 721tx, Mz7, N. H. Feniak, Nick Number, Nuno Tavares, O'Dubhghaill, Oshwah, P.Geol, Paul H., Peter M. Brown, Phe, Philip72, Pldms, Pythoncoder, Quiddity, Rich Farmbrough, RockMagnetist, SD5, Shellwood, Siim, Simone, Smith609, Stephenb, Supermax2424, TGIP KING, TaBOT-zerem, The PIPE, Thingg, Tom Radulovich, UY Scuti, Vipinhari, Vsmith, Vuong Ngan Ha, WadeSimMiser, Waso99, Whatiguana, Woodlot, Yamara, Yrithinnd, Zedshort, Zzyzx11, 팔레오세와 에오세, 107 anonymous edits 241

Paleocene *Source:* https://en.wikipedia.org/w/index.php?oldid=852406425 *License:* Creative Commons Attribution-Share Alike 3.0 *Contributors:* 5 albert square, A3RO, AMK152, Aboalbiss, Abyssal, Alfie Gandon, Alphathon, Antandrus, Anthony Appleyard, Apidium23, Apokryltaros, Arglebargle79, Aristox, Ballista, Bejnar, BenoniBot~enwiki, Bill, Blueshifter, Bluesmoon, Bob the Wikipedian, Bobo The Ninja, Bobo192, Butko, Cabrochu, Cadiomals, CambridgeBayWeather, Canthusus, Casliber, Chaoyangopterus, Chermundy, Chris the speller, ChrisGualtieri, Civil Engineer III, ClueBot NG, CommonsDelinker, Courcelles, Crocadog, Cwobeel, DVD R W, DadaNeem, Davidiad, Dawnseeker2000, December21st2012Freak, Deflective, Deor, Dinoguy2, Download, Dysepsion, Dysmorodrepanis~enwiki, Edderso, Environnement2100, Eperotao, Eras-mus, Erimus, Excirial, Fafnir1, Falconfly, Fama Clamosa, Firsfron, Fossiliferous, Freemo1999, Gap9551, GeoWriter, Geologyguy, Georgialh, Gilgamesh~enwiki, Glenn, Gob Lofa, Gorthian, HMSSolent, Headbomb, Iblardi, Indon, Ipele222, IvanLanin, J. 'mach' wust, J. Spencer, J.delanoy, Janet1983, Jer ome, Jinfengopteryx, John Reaves, JorisvS, KP Botany, Kid2, Kingpin13, Kwamikagami, Kyzul2, Leptictidium, Leszek Jańczuk, LilHelpa, Look2See1, Lotje, M0rphzone, Marcelo-Silva, Markjoseph125, Mattmujica, Metanoid, Mgiganteus1, Mhese, Mikenorton, Mikeo, Minigirl7, Mojoworker, NatureA16, No Swan So Fine, Numbo3, Omnipaedista, OrinR, Oshwah, Oxymoron83, P.Geol, PainMan, PatternSpider, Peter M. Brown, PlyrStar93, Pterre, Puchiko, Pukaluka, Quentar~enwiki, RL0919, Rjwilmsi, Ron Ritzman, Ruakh, Saigon punkid, SamX, Shellwood, Sinn, SkyLined, Slate Weasel, Smithsonwick, Squidboi010, Steinbach, Steven Crossin, Sunomi64, Thanatos666, The Thing That Should Not Be, Tianasez, Tinomac, Tnxman307, Tobias Paul, Tsinoyman, Underalms, UpdateNerd, VMS Mosaic, Vanamonde93, Versus22, Vsmith, Warfreak, Wesley J M, Wetman, Wikimedes, WolfmanSF, WolfyB, Woudloper, Xenonice, Yaziel, ماني, 181 anonymous edits 245

Eocene *Source:* https://en.wikipedia.org/w/index.php?oldid=853513289 *License:* Creative Commons Attribution-Share Alike 3.0 *Contributors:* 37ophiuchi, A. Parrot, AManWithNoPlan, Abyssal, Alan Liefting, Altenmann, Anaxial, Andrewjlockley, Apokryltaros, Apollonius 1236, Arado, Armascout, Arthena, Awickert, Axeman89, Azollamfc, BD2412, BOT-Superzerocool, Bender235, Bento00, BlaiseFEgan, Bobrayner, Bridgetswade, Brothernight, Butko, Cadiomals, Calaschysm, Centibyte, Ched, Chermundy, Citation bot 2, ClueBot NG, CommonsDelinker, Cyrus noto3at bulaga, Davidiad, Dcirovic, Deflective, Detect~enwiki, Editor1023, Elisevil, ErikHaugen, Erimus, Explicit, Fafnir1, Fama Clamosa, Fangorn-Y, Fedginator, Flyer22 Reborn, Fotaun, FunkMonk, GVP Webmaster, GeoWriter, Geologyguy, Gilgamesh~enwiki, Gilliam, Gnostic804, Gob Lofa, Gug01, Gulumeemee, GünniX, I dream of horses, Invertzoo, IronGargoyle, J. 'mach' wust, JD7788, JFG, JLaTondre, Jamesinderbyshire, Jdawg123454123, Jerryrd, Jillswin, Jj dowg, John Troodon, JohnSka, Jojhutton, Joseph Solis in Australia, Jusdafax, Karath88, Karshan, Kevmin, Kits1952, Kwamikagami, Kyzul, Leptictidium, Leszek Jańczuk, Logan, Look2See1, Maias, Malurian123, MangoWong, Materialscientist, MaynardClark, Meganesia, Mejor Los Indios, Mhese, Mikenorton, Morbas, Muddybootz, MusikAnimal, Nathan Johnson, NatureA16, Niceguyedc, Noformation, Noobeditor, NottNott, ObersterGenosse, Omnipaedista, Orangemarlin, Orenburg1, Oshwah, Paul H., Peter M. Brown, Petrb, Phe, Pmokeefe, Pterre, Ran, Reach Out to the Truth, Rene Sylvestersen, Rich Farmbrough, Richard Keatinge, Rock4arolla, RockMagnetist, Rudolf Pohl, Rursus, Rwflammang, Ryan Vesey, Sierra Shaver, SkyLined, Smith609, Soap, Spicemix, Stealth cat, Stephen MUFC, Storkk, Thanatos666, TheEpTic, Thingg, Tide rolls, Uhai, VexorAbVikipædia, Vkmuliya, VoABot II, Vsmith, Wayne Hardman, Websterwebfoot, Wenkbrauwalbatros, WereSpielChequers, Whoop whoop pull up, Widr, William M. Connolley, Wilson44691, WolfmanSF, Wsanders, Xook1kai Choa6aur, Zamphuor, Zocke1r, דקי', 162 anonymous edits 252

Oligocene *Source:* https://en.wikipedia.org/w/index.php?oldid=847487288 *License:* Creative Commons Attribution-Share Alike 3.0 *Contributors:* 37ophiuchi, Abyssal, Alansohn, Alexf, Anaxial, Andrea105, Animalparty, Apidium23, Arch dude, BD2412, Bobo192, Bobrayner, Bomac, Brian Crawford, Brothernight, Byteflush, CAPTAIN RAJU, Cadiomals, Chermundy, Chingchuck13, Chris the speller, ClueBot NG, Cnwilliams, CommonsDelinker, Courcelles, Dana boomer, Danielklein, Darth Panda, Dave.Dunford, Davidiad, Dawnseeker2000, Deflective, Dicklyon, Discospinster, DrFO.Jr.Tn~enwiki, DuncanHill, Enlil Ninlil, Ericl, Erimus, Excedron9238, Fafnir1, Falconfly, FunkMonk, GENVELES, Gail, Gap9551, GeoWriter, Gob Lofa, Graham87, HalfShadow, HappyInGeneral, IronGargoyle, J. 'mach' wust, J. Spencer, Jackakraw, Jackhynes, Jbeans, Jonesey95, JorisvS, Jovianeye, Koebie, Kwamikagami, Leptictidium, Look2See1, MC10, Marek69, Martijn faassen, Mejor Los Indios, Mercurywoodrose, Metalraptor, Mhese, Mikenorton, Morbas, Nihiltres, Nomadic Whitt, Nonsenseferret, Obsidian Soul, Olev Vinn, Oliver202, Omnipaedista, Orangemarlin, PageRob, Parsa, Peter M. Brown, Philip Trueman, Pinethicket, Pixor, Rcsprinter123, Reedy, Rich Farmbrough, Richard New Forest, Rjwilmsi, RockMagnetist, SamX, Seasonal Note, SkyLined, Slightsmile, Spencer62, StarryGrandma, Stealth cat, Swid, Tempodivalse, Thanatos666, Tommy2010, Trappist the monk, Trevor MacInnis, Vsmith, WereSpielChequers, Wetman, Widefox, Widr, William M. Connolley, Willking1979, Wjejskenewr, Wutsje, Z0, 133 anonymous edits 266

Neogene *Source:* https://en.wikipedia.org/w/index.php?oldid=853513776 *License:* Creative Commons Attribution-Share Alike 3.0 *Contributors:* 37ophiuchi, A2-33, AMK152, Abyssal, AlefZet, Alfie Gandon, Andycjp, Arkuat, Arnon Chaffin, Avoided, Awickert, Bejnar, Bryan Derksen, CLCStudent, Cadiomals, Calvinteah96, Capricorn42, Casito, Chermundy, ClueBot NG, CommonsDelinker, Conversion script, Creeper122, DHN-bot~enwiki, Danish Pig Margaret, Dawnseeker2000, Deflective, Denton22, Derek R Bullamore, Derek Ross, Discospinster, Djhmoore, Dom Damian, Don Kenney, Downwards, DrewCR7, Drewmutt, Drmies, DuncanHill, Dysmorodrepanis~enwiki, Ekren, El C, Emperorbma, Epbr123, Eusis, FaerieInGrey, Fielddaysunday, Finnusertop, Gaius Cornelius, Gap9551, GeoWriter, Geologyguy, Glenn, Goodvac, Graham87, Gurch, Hbf1184, Hike395, Hko2333, Infrogmation,

354

Image Sources, Licenses and Contributors

The sources listed for each image provide more detailed licensing information including the copyright status, the copyright owner, and the license conditions.

358

License

Index

Australasia, 11
Australia, 12, 13, 63, 126, 178, 186, 219, 246, 247, 259, 261, 284, 286, 321
Australia (continent), 239
Australopithecine, 297, 303
Australopithecus, 59, 236
Australopithecus garhi, 323
Avalonia, 6, 91, 112
Aves, 208, 263
Avialan, 175, 211
Avicephala, 165
Aviculopecten, 146
Azolla, 254, 255, 258
Azolla event, 239, 254, 255, 258

Bacteria, 21, 37, 131, 142
Bactritida, 123
Baculites, 225
Bajocian, 202
Baleen whale, 271
Baltica, 5, 6, 8, 26, 50, 68, 76, 91, 112, 124, 140
Baltic amber, 263
Baltic Sea, 16, 237, 263, 310, 329
Baltic Shield, 34
Baltoscandia, 76
Banded iron formation, 4, 24, 26, 34, 35, 41
Baragwanathia, 111, 114
Bark (biology), 141
Barremian, 217
Barstovian, 283
Barstow Formation, 287
Bartonian, 245, 252, 253, 266
Baryte, 37
Basal (phylogenetics), 151, 191, 211
Basalt, 35, 42
Bashkirian, 137, 138
Basidiomycetes, 142
Basilosaurus, 57, 58, 235, 238, 263
Bat, 240, 261
Batholith, 11, 203
Bathonian, 202
Batoidea, 225
Batrachosauria, 165
Bats, 289
BBC, 17, 85
Bear, 286
Beaver, 286
Bee, 221
Beech, 268
Beestonian stage, 294, 318
Before present, 233, 292, 309, 313, 320, 325
Behemotops, 271
Belemnite, 219
Belemnitida, 206
Bembridge Marls, 263

Bendigo stage, 89
Bennettitales, 187, 208, 211, 221
Beringia, 296, 310
Bering Strait, 310
Berriasian, 217
Berruornis, 251
Bibcode, 27, 32, 61, 84, 170, 213, 232
Bibionidae, 263
Billion years, 34
Biocoenosis, 229
Biodiversity, 61, 228
Bioerosion, 96, 206, 225, 230
Biogenic substance, 37
Biomass, 181
Biostratigraphy, 21, 85, 309
Biotic material, 30
Bioturbation, 78
Bipedalism, 287
Bird, 154, 172, 181, 201, 208, 222, 240, 248, 278, 279, 289
Birds, 216, 244
Bison, 279
Bivalve, 92, 96, 146, 147, 150, 207
Bivalves, 271
Bivalvia, 206
Black Hills stage, 89
Black Sea, 236, 310
Blancan, 294, 322
Blastoid, 146
Blattoptera, 164
Blytt–Sernander system, 309, 313, 325, 327
Bohemia, 110
Bolide, 197, 253
Bolide impact, 331
Bolinda stage, 89
Bømlo, 186
Bond albedo, 36
Bond events, 330
Boreal (age), 325
Borealosuchus, 250
Boreal (period), 309, 313, 325, 326
Boreas (journal), 312
Borhyaenidae, 240, 269
Borophaginae, 286, 303
Bosphorus, 310
Bothriolepis, 127
Brachiopod, 93, 96, 100, 121, 125, 128, 145, 146, 148, 162, 196
Brachiopoda, 95, 142
Brachiopods, 105, 114
Brachiosaurus, 54, 174, 208
Braided river, 317
Brain to body mass ratio, 249
Bramertonian Stage, 294
Branchiosauridae, 189
Brasilitherium, 187

Brasilodon, 187
Brazil, 34, 187
British English, 242
Brodia, 151
Brontothere, 240
Brontotheriidae, 269
Bronze Age, 327
Brooks Range, 177
Brown alga, 288
Browsing (herbivory), 247
Bryophyte, 104, 109
Bryozoa, 78, 93, 96, 100, 114, 125, 145, 146
Bryozoan, 97, 102
Bryozoans, 105
Burdigalian, 278, 281, 283, 292
Burgess Shale, 22, 79, 84
Burrell stage, 89
Butterfly, 223
Bya, 34

C3 carbon fixation, 285
Cactus, 247
Cadastre, 111
Caecilian, 208
Caiman, 288
Cainozoic, 241
Calabrian (stage), 309, 313, 325
Calamites, 145
Calcareous, 225
Calcite, 203, 217, 320
Calcite sea, 92, 203
Calcium carbonate, 95, 216
Caledonian Mountains, 7
Caledonian orogeny, 112, 113, 124
Callistophytales, 144
Callovian, 202
Camarasaurus, 208
Cambria, 20, 73
Cambrian, 5, 20, 25, 43, 46, 48, 62, 63, 68, 71,
    71, 84, 86, 87, 95, 96, 104, 107, 110,
    119, 131, 135, 158, 168, 182, 199, 214,
    241, 276, 308
Cambrian explosion, 22, 48, 63, 64, 73, 95
Cambrian–Ordovician extinction event, 6, 88
Cambrian Series 2, 75
Cambrian substrate revolution, 78
Cambridge University, 306
Cambridge University Press, 198, 349
Camel, 303, 321
Camelid, 269, 286
Camelops, 310
Campanian, 217, 221
Canada, 11, 30, 197
Canadians, 133
Canadian Shield, 23, 34, 310
Candona, 150

Canid, 271, 286
Caninia (genus), 146
Capitanian, 160
Capitosaur, 190
Captorhinidae, 154
Caradoc Series, 89, 90
Carbon, 21, 142, 260
Carbon-12, 252
Carbon-13, 75, 252
Carbon-14, 315
Carbonate, 12, 34, 126, 139, 220
Carbonate hardgrounds, 92, 96, 203
Carbon burial, 41
Carbon dioxide, 31, 34, 68, 71, 86, 107, 119,
    132, 135, 158, 178, 182, 199, 214, 221,
    241, 253, 260, 276, 285, 308
Carbon dioxide in Earths atmosphere, 318
Carbonicola, 146, 150
Carboniferous, 5, 7, 48, 62, 63, 68, 71, 86, 107,
    119–121, 135, **135**, 137, 157–159, 162,
    164, 178, 182, 189, 199, 214, 241, 276,
    308
Carboniferous Limestone, 138
Carboniferous Rainforest Collapse, 52, 66, 69,
    137, 141, 154, 156, 160
Carboniferous System, 157
Carbon isotope, 252, 258
Carbonita (genus), 150
Carbon sequestration, 285
Carbon sink, 132
Carcharhinid, 247, 271
Carcharhiniformes, 263
Carcharocles chubutensis, 288
Carcharodontosaurus, 55, 175
Cardiocarpus, 145
Carmarthenshire, 110
Carnian, 184, 191
Carnian Pluvial Event, 178, 187
Carnivora, 235
Carnivore, 165, 180, 229
Carnivores, 303
Carrion, 229
Cashew, 268
Caspian Sea, 236
Castlemaine stage, 89
Castorocauda, 174
Catkin, 145
Caturrita Formation, 187
Caulopteris, 145
Cautley stage, 89
Cave, 323
Cave bear, 321
Caviomorpha, 304
Caytoniaceae, 211
Cayugan, 111
Celsius, 178

Giant ground sloth, 304
Giant ground sloths, 60, 237
Giant impact hypothesis, 3, 21, 26, 31
Giant sloth, 321, 331
Gideon Mantell, 171
Gigaannum, 30
Giga-annum, 26
Gigantism, 142
Gigantophis, 250
Gigantopithecus, 321
Gigantoproductus, 146
Gill, 114
Ginkgo, 162, 164, 179, 187, 211
Ginkgo biloba, 179
Ginkgophyta, 187
Giovanni Arduino (geologist), 308
Giraffe, 303
Giraffid, 321
Gisborne stage, 89
Givetian, 123
Glacial, 309, 313, 325
Glacial period, 282
Glacial striation, 163
Glaciation, 23, 137, 139, 156, 285, 313, 315
Glaciations, 59, 237
Glacier, 11, 15, 40, 63, 91, 93, 112, 113, 124,
    139, 178, 187, 202, 294, 308
Global Boundary Stratotype Section and Point,
    25, 74, 315, 346
Global warming potential, 254
Glomerales, 104
Glomerula, 206
Glossopteridales, 187
Glossopteris, 164, 187, 194
Glyptodon, 310, 321, 322
Glyptodont, 303, 304
Gnathostomata, 88, 109, 123
Gneiss, 36
Göbekli Tepe, 331
Gobiconodon, 176
Gomphothere, 303, 321
Gomphotheres, 286
Gondwana, 5–7, 9, 10, 12, 13, 42, 48, 50, 52,
    54, 63, 64, 66, 68, 76, 88, 91, 93, 105,
    111, 112, 121, 124, 139, 140, 162, 169,
    177, 178, 184, 191, 201, 202, 219
Gondwanaland, 247
Gondwanathere, 177, 269, 287
Goniatite, 123, 146
Gorgonopsia, 67, 165, 180
Gorgonopsid, 52
Grande Coupure, 236, 266
Granite, 34
Granitic, 34
Graphite, 37
Graptolite, 88, 93, 96, 105

Grass, 236, 240, 268
Grasses, 261, 285
Grasshopper, 223
Grassland, 266, 282, 285, 297
Graywacke, 4, 34
Grazers, 285
Great American Interchange, 296, 297
Great Britain, 7, 124, 294, 317
Great Dying, 172
Great Lakes, 59, 237, 310, 311
Great Ordovician Biodiversification Event, 78,
    88
Great Oxygenation Event, 23, 40
Great Plains, 284
Great Salt Lake, 317
Grebe, 225
Greek language, 234, 314
Green algae, 79, 104
Greenhouse, 6, 105, 113
Greenhouse gas, 260, 285
Greenhouse gases, 169
Greenland, 13, 21, 30, 34, 221, 246, 260, 272,
    285
Greenlandian, 309, 313, 325, 326
Greenland ice sheet, 294
Green River Formation, 260
Greenstone belt, 4, 34
Grenville orogeny, 42
Ground sloth, 303, 321
Group (stratigraphy), 219
GSSP, 309
Guiana Shield, 31
Gulf of Mexico, 10, 202, 216, 244
Gulf Stream, 239, 278
Gunz glaciation, 318
Gymnosperm, 144, 162, 179, 211, 221
Gyracanthus, 152
Gzhelian, 137, 138

Haasts eagle, 321
Hadean, 25, **27**, 34, 37, 47
Hadean zircon, 31
Hades, 30
Hadley cell, 124
Hadrosaurs, 56, 176
Hakel, 227
Halite, 178
Hallucigenia, 83
Hamburg culture, 331
Hand, 261
Hardground, 99, 230
Hardgrounds, 225
Haughton impact crater, 274
Hauterivian, 217
Heat, 259
Hederellid, 114, 125, 145

www.ingramcontent.com/pod-product-compliance
Lightning Source LLC
Chambersburg PA
CBHW031040110426
42740CB00047B/765